MW00999836

First published in 2017 by Voyageur Press, an imprint of The Quarto Group, 401 Second Avenue North, Suite 310, Minneapolis, MN 55401 USA. Telephone: (612) 344-8100 Fax: (612) 344-8692

QuartoKnows.com
Visit our blogs at QuartoKnows.com

10 9 8 7 6 5 4 3 2

ISBN: 978-0-7603-5264-9

Library of Congress Cataloging-in-Publication Data

Names: Larsen, Jon, 1959-
Title: In search of stardust : amazing micrometeorites and their terrestrial imposters / Jon Larsen.
Other titles: In search of star dust
Description: Minneapolis, Minnesota : Voyageur Press, an imprint of Quarto Publishing Group, [2017]
Identifiers: LCCN 2016057352 | ISBN 9780760352649 (hc)
Subjects: LCSH: Meteorites--Identification. | Rocks--Identification.
Classification: LCC QB755 .L37 2017 | DDC 523.5/1--dc23
LC record available at https://lccn.loc.gov/2016057352

ACQUIRING EDITOR: Dennis Pernu
PROJECT MANAGER: Jordan Wiklund
ART DIRECTOR: Cindy Samargia Laun
DESIGN AND LAYOUT: Kjell Olufsen
ADDITIONAL LAYOUT: Rebecca Pagel

Printed in China

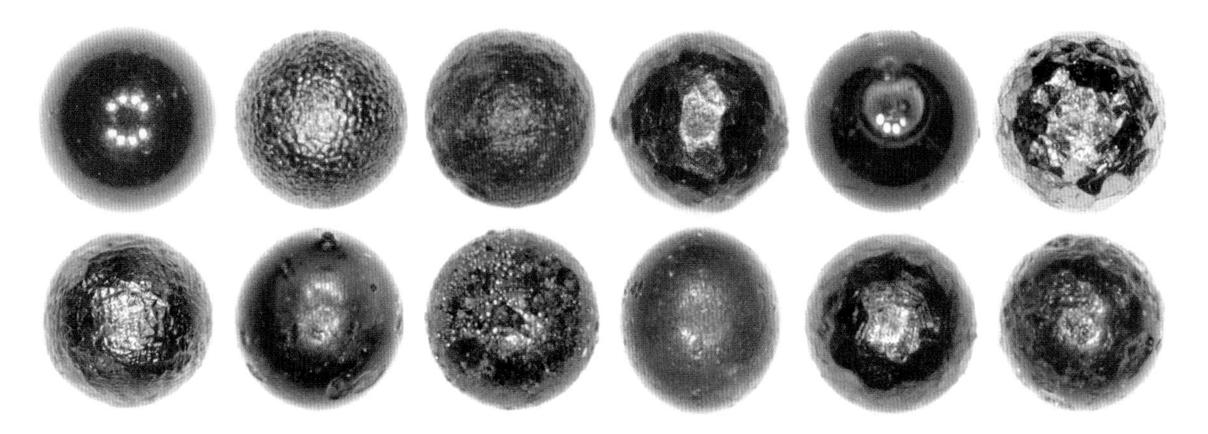

IN SEARCH OF STARDUST

Amazing Micrometeorites and Their Terrestrial Imposters

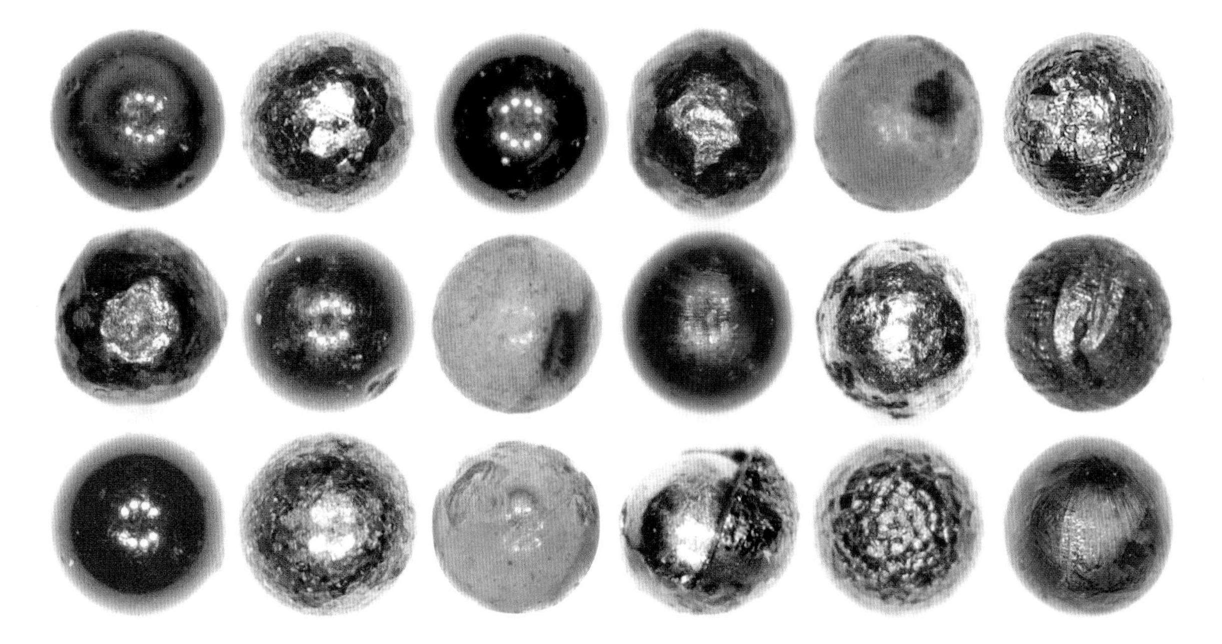

Jon Larsen

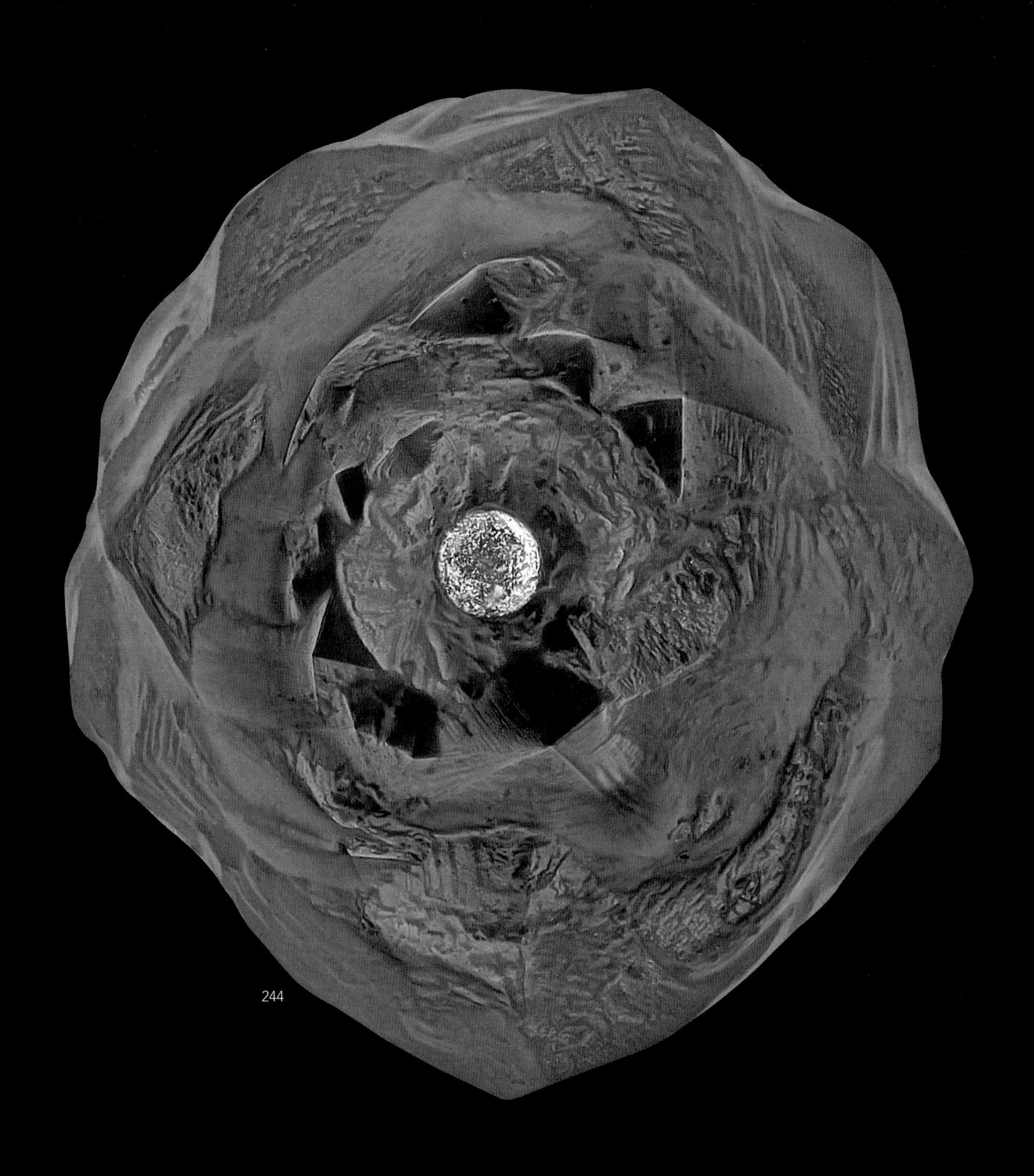

244

Contents

OPPOSITE: A fine-grained *turtleback* micrometeorite with a nickel/iron core in the front, revealed by the ablation during the atmospheric flight.

72

Preface

Is it possible to find micrometeorites in populated areas?

The question has been raised for nearly a century, and despite numerous attempts to find them, the answer up to this day has been a very short *no*. This was confirmed by the top researchers in the field when my project started up.

Meanwhile our knowledge about these amazing stones has gradually increased. There is a continuous evolutionary line in the research on micrometeorites, from the early pioneers John Murray and Adolf Erik Nordenskiöld to Lucien Rudaux and Harvey H. Nininger. With Donald E. Brownlee and Michel Maurette in the '60s micrometeoritics became real science. During the past two decades this research has accelerated, thanks to, among others, Susan Taylor who extracted micrometeorites from the South Pole water well, and Matthew Genge who figured out the splendid classificiation. Today there is a growing literature about micrometeorites, but still the answer to the initial question has been no.

Micrometeorites have mainly been found in the Antarctic, but also to some extent in prehistoric sediments, remote deserts, and in glaciers—places that are clear of the confusing anthropogenic influence. In populated areas a wall of contamination has been considered insurmountable.

It is therefore with pride and joy I can report here about a project of systematic examination of all sorts of anthropogenic and naturally occurring spherules in an empirical search for micrometeorites in populated areas. This research has resulted in a new collection of pristine cosmic spherules. The findings have been analysed at several different institutions including electron microprobe verification at the Natural History Museum in London. This new collection of urban micrometeorites is presented here for the first time.

Without knowing what micro-meteorites really look like, it would not have been possible to find them, and it is a pleasure also to present here for the very first time a morphological study of micrometeorites in high-resolution color photography. This has become possible thanks to new micro photo techniques developed together with my brilliant colleague Jan Braly Kihle—*sine qua non*!

Furthermore, this research would have been unsuccessful if not for the invaluable support from the Laboratory for Electron Microscopy in Bergen (UiB), Egil Severin Erichsen, Irene Heggstad, and Gunnar Sælen, the SEM lab at the University in Oslo (UiO) Berit Løken Berg and Henning Dypvik and the Natural History Museum (NHM) in Oslo, Rune Selbekk and Harald Folvik—warm thanks to these splendid fellow researchers. But above all very special thanks to Matthew Genge at Imperial College, London, who not only verified my first micro-meteorites but also has been my mentor *pro bono,* and who initiated the crucial electron microprobe analysis of the micrometeorites at the NHM in London.

Jon Larsen

OPPOSITE: SEM images on pages 14–21 are mainly from the analysis of the new collection performed at the Natural History Museum (NHM), London, by Matthew Genge and the author. The catalog number accompanies each stone. Some of the images on the pages 14–21 have a catalog number with a hyphen. These are from the South Pole Water Well (SPWW), the Antarctic reference collection of micrometeorites, SEM images by Emily Schaller, published with kind permission from Susan Taylor (US Army Cold Regions Research, CRR).

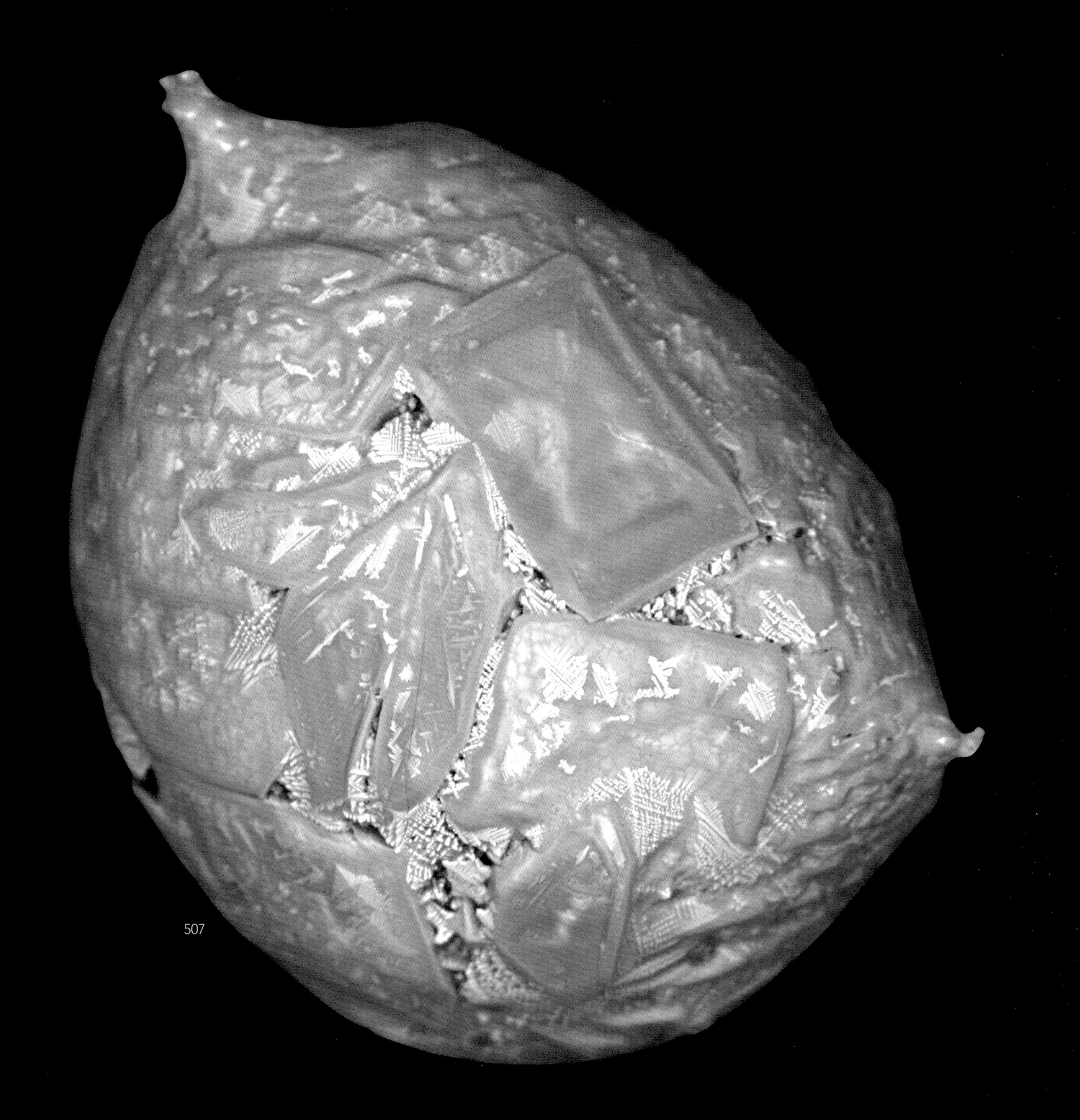

507

The Stardust Project

Micrometeorites belong to the oldest matter there is: mineral remnants from before the planets were formed. They may even contain stardust older than the sun, particles which have traveled farther than anything else on Earth. We are just beginning to explore these alien stones, yet they are everywhere around us.

After an incident in 2009 when a micrometeorite literally landed on my table, I wanted to find out more about them, and became intrigued by the contradiction between the global influx rate (see page 14) and the established postulate that they could not be found in populated areas. Thanks to Gunnar Sælen at the Bergen University I got access to the library and a growing number of academic publications on the subject. It soon became clear that the problem was the unexplored wall of anthropogenic contaminants. As we shall see from page 82 in this book, human tools and activities create spherules not unlike micrometeorites.

To pick out one extraterrestrial particle among billions of others requires knowledge both about what to look for and what to disregard, and initially I was in the dark. The published images of micrometeorites from the Antarctic were mainly black/white SEM sections (pages 16–23), which are poor representations of what micrometeorites look like. And with regard to the confusing contaminants there was a plentitude of speculation, but very little empirical data. There has been research in this field earlier, from NASA's comparative analysis of cosmic and industrial spherules in the 1960s to contemporary studies of road dust in India and Hungary, but these have been fragmentary, and concluded that separation of micrometeorites in populated areas is not possible. On the other hand there have been several physics-at-home experiments searching for micrometeorites at downspouts from roofs gutters, but none of these have yet resulted in verification of extraterrestrial particles.

In the spring of 2010 I started my systematic research on dust samples from populated areas. I initially looked at skywards-facing hard surfaces where particles could accumulate over time, like roads, roofs, parking lots, and industrial areas, and then graduated to look in other cities, countries, mountains, beach sand, deserts—everywhere. Now, six years later, I can look back upon nearly 1,000 field searches in nearly 50 countries, all continents represented. The samples were examined in a Zeiss binocular microscope, and interesting particles were picked out, photographed with a USB microscope, and stored in the archive. I established a photo database (now containing photos of more than 40,000 individual objects), kept an illustrated journal, and tried to look for patterns (factor analysis) while I put my complete trust in pure empiricism. The Facebook page *Project Stardust* was established to share the results.

To begin with the different types of anthropogenic and naturally occurring terrestrial spherules seemed infinite and chaotic, but with time I gradually started to recognise the most common ones. There are surprisingly small variations in the types of spherules found in comparable environments around the globe, and the 25 types presented in this book represent the most of all the spherules found anywhere. The micrometeorites are rare and evenly distributed, so

Origin, Formation, Influx, and Classification

There are almost as many explanations as to where the micrometeorites have their origin as there are researchers in the field. Depending upon who you ask, the answer may vary from the asteroid belt between Mars and Jupiter, comet-related objects in the Kuiper belt or Oort cloud, various planetary ejecta, interstellar matter, and so on. It is estimated that up to 0.1% of the matter in primitive meteorites, possibly also in micrometeorites, are presolar grains. On the other hand there are achondritic (igneous) micrometeorites from differentiated bodies like the Moon and Vesta. Throughout history large asteroid impacts on the rocky planets and their moons have ejected substantial quantities of rocks into space, and it is possible to imagine an extensive exchange of matter between all the planetary bodies and their surrounding dust rings, with the zodiac cloud as a temporary storage pool. This was considered sci-fi only a few years ago.

Hopefully access to a new, potentially large source of micrometeorites in populated areas may contribute to a systematic mapping of the isotopic variations of a substantial number of micrometeorites in the years to come, with more data about the micrometeorites parent bodies as a result. It should not come as a surprise if the origins of the micrometeorites turn out to be a combination of all dust-producing bodies in the solar system and beyond.

The micrometeoroids enter the Earth's atmosphere with a speed up to 50 times that of a rifle bullet. Depending upon the entry angle (see page 15) relative to Earth's rotation, the peak temperature from the frictional heat will cause a substantial variation in the alteration process. Approximately half of the micrometeoroids <0.1 mm receive a soft deceleration and end on the ground as unmelted micrometeorites. The other half reach peak temperatures between 2,460–3,630°F (1,350–2,000°C), which is enough to create the various types of melted cosmic spherules

~ 2,462 °F / 1,350 °C

2,732 °F / 1,500 °C

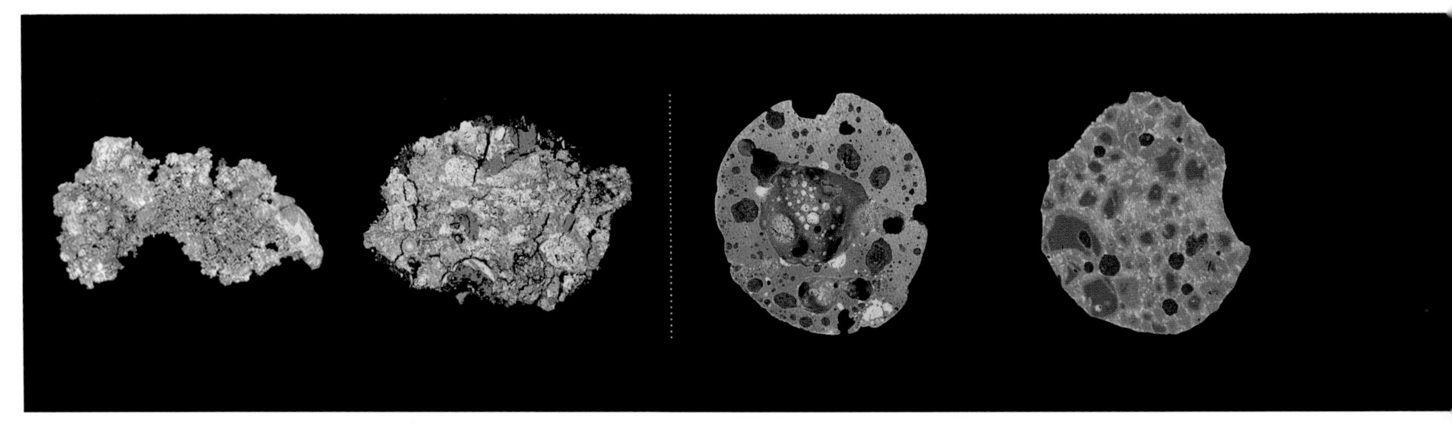

UNMELTED

SCORIACEOUS

(see illustration on these two page). A sphere is nature's solution to maximum volume with the smallest possible surface, and is created by the surface tension in a liquid state. At the same time a rapid differentiation takes place, where the heavier elements (iron, nickel, platinum, etc.) move inwards to form a core, and volatile elements escape. Iron from the stone reacts with oxygen in the atmosphere and creates dendritic magnetite, looking like small Christmas trees on the surface. Still in flight but decelerating, the inertia of the heavy core may push it forward in the direction of travel, often spinning, while the ablation erodes characteristic aerodynamic forms. The whole formation is over in the blink of an eye before the micrometeoroids fall to Earth at terminal velocity. Based upon radar measurements the general influx rate of micrometeorites is estimated to approximately one object with a diameter 0.1 mm per square meter per year.

The *Classification of Micrometeorites* was published in 2008 by Matthew Genge together with Cécile Engrand, Matthieu Gounelle, and Susan Taylor, and is the most comprehensive article about micrometeorites to date. It is freely available on the Internet, and is a must for everyone interested in micrometeorites. As explained in the previous paragraph, the various types of micrometeorites are mainly formed by the peak temperature in combination with the quenching profile (fast/slow cooling) during the atmospheric flight. There are transitional forms between the various types, but the chemistry of the micrometeorites is surprisingly homogenous, mainly chondritic (see spectrum on page 11), but with some minor (or rare) variations. Future research may add varieties to the present classification, and with more hands and eyes in the field, micrometeoritics can evolve into an exciting new branch of the popular study of space rocks—ever expanding, the same as the universe itself.

2,912 °F / 1,600 °C | 3,272°F / 1,800 °C | 3,452 °F / 1,900 °C | 3,632 °F / 2,000 °C

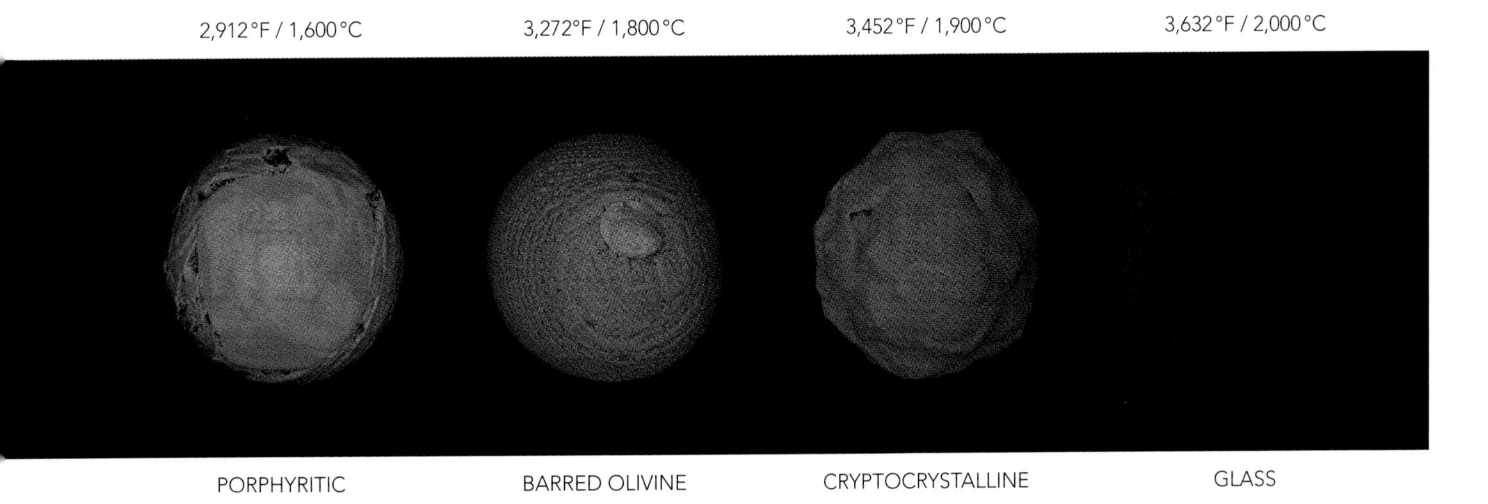

PORPHYRITIC | BARRED OLIVINE | CRYPTOCRYSTALLINE | GLASS

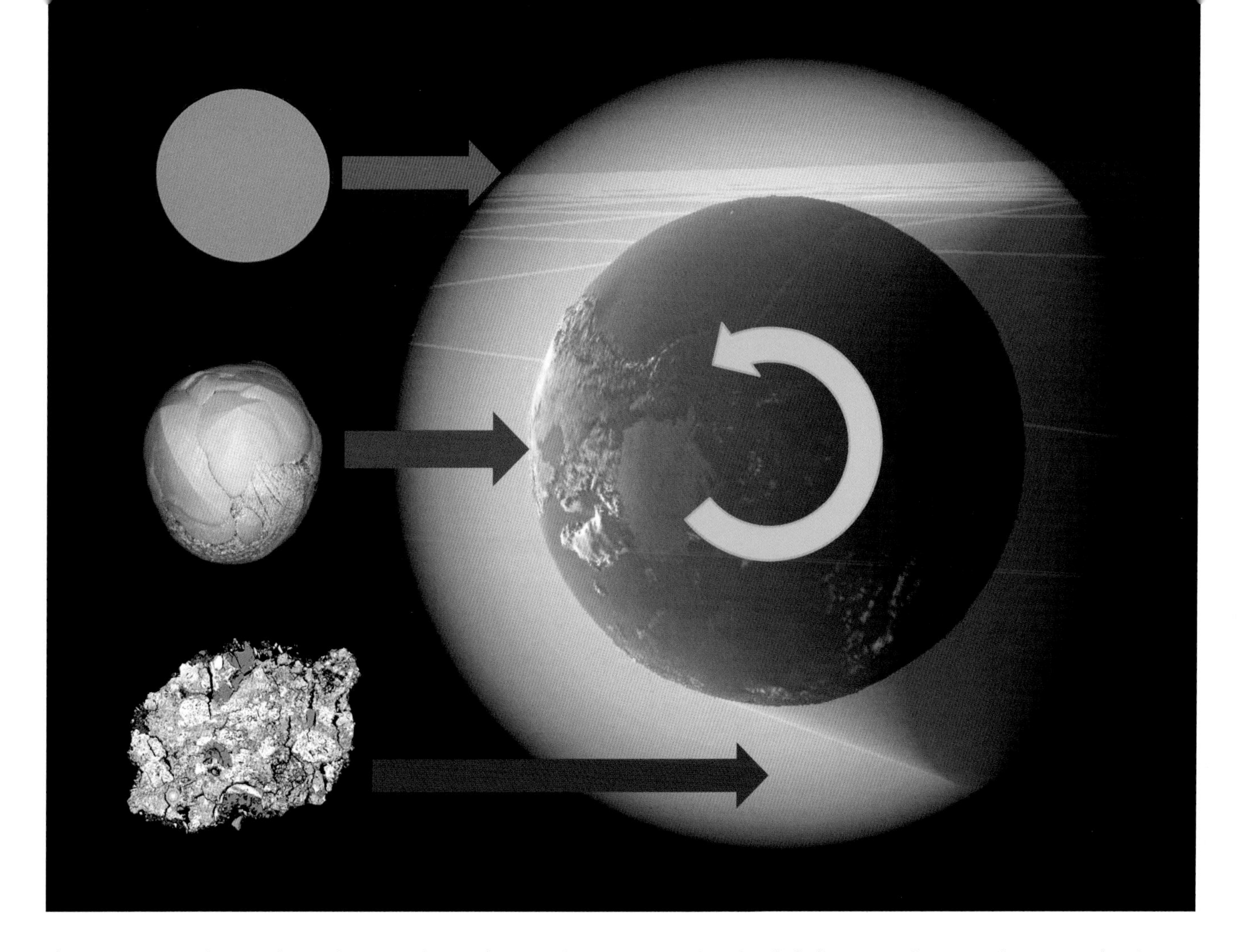

The micrometeoroids enter the Earth's atmosphere with a speed up to 50 times that of a rifle bullet. Depending upon the entry angle relative to Earth's rotation, the peak temperature from the frictional heat will cause a substantial variation in the formation process.

In the illustration on this page we see a completely melted (glassy) micrometeorite on top, a fine-grained/barred olivine combi-micrometeorite (middle), and an unmelted micrometeorite (bottom). Approximately half of the micrometeoroids less than 0.1 mm receive a soft deceleration and end on the ground as unmelted micrometeorites. The other half reach peak temperatures between 2,460–3,630°F (1,350–2,000°C), which is enough to create the various types of melted cosmic spherules.

OPPOSITE: Mass distribution of the extraterrestrial influx on Earth. To the left are the micrometeorites, with a distinct peak between 0.2 and 0.4 mm, before it decreases to zero between approximately 2 millimeter and 1 centimeter, this is where the meteors appear. With larger mass and kinetic energy they burn up in the atmosphere, leaving behind only nano-sized meteoritic smoke particles. At around 1 centimeter and upwards to a few meters are the meteorites, and to the right are the large, but rare asteroids.

The micrometeorites is the largest of the three groups of extraterrestrial matter, and the influx rate is one object with a diameter of app. 0.1 mm per square meter per year. However, the average cosmic spherule has a diameter around 0.3 mm, which contains up to 27 times more matter than one at 0.1 mm. Consequently, in our hunt for MMs in populated areas, on a roof of 50 square meters, we can expect to find 2, and not 50 cosmic spherules per year.

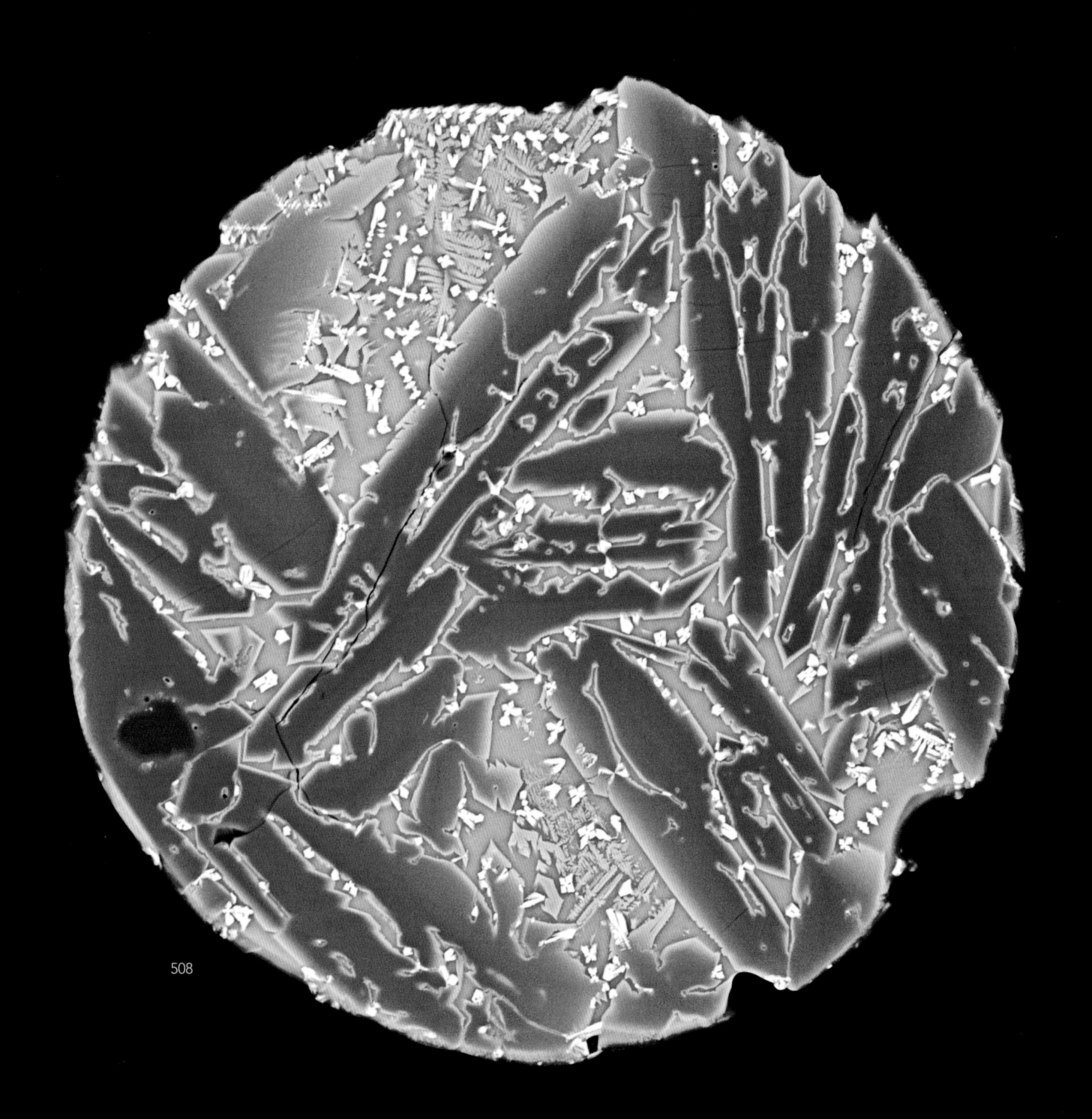

MICROMETEORITES

Scanning-Electron Micrographs

Scanning-electron microscope (SEM) section images of micrometeorites molded in resin enable precise analysis of the inner structure of the stones but give little clue about what they really look like. It is, however, necessary to know the basic structure of the micrometeorites in order to understand their morphology and ultimately to be able to recognize them in dust samples from populated areas.

The backscatter SEM images on pages 16–23 are mainly from the analysis of the new collection performed at the Natural History Museum (NHM), London, by Matthew Genge and the author. The catalog number follows each stone, and for the first time enables comparison between the color photos and the SEM images. On page 68 are three micrometeorites, each depicted in three different ways: color photo, SEM surface image, and SEM section image.

Some of the images on the pages 16–23 have a catalog number with a hyphen. These are from the South Pole Water Well (SPWW), the Antarctic reference collection of micrometeorites, SEM images by Emily Schaller, Dartmouth College, published with kind permission from Susan Taylor (US Army Cold Regions Research, CRR).

The SEM section image to the left is a porphyritic micrometeorite (~0.2 mm) with large, dark olivine crystals, gray dendritic pyroxene, and white specks of magnetite crystals. Below are three distinct types of micrometeorites, from left: porphyritic (see pages 19, 21, and 35-37), cryptocrystalline (a turtleback, pages 4, 27, and 42) and barred olivine (pages 18, 38-41).

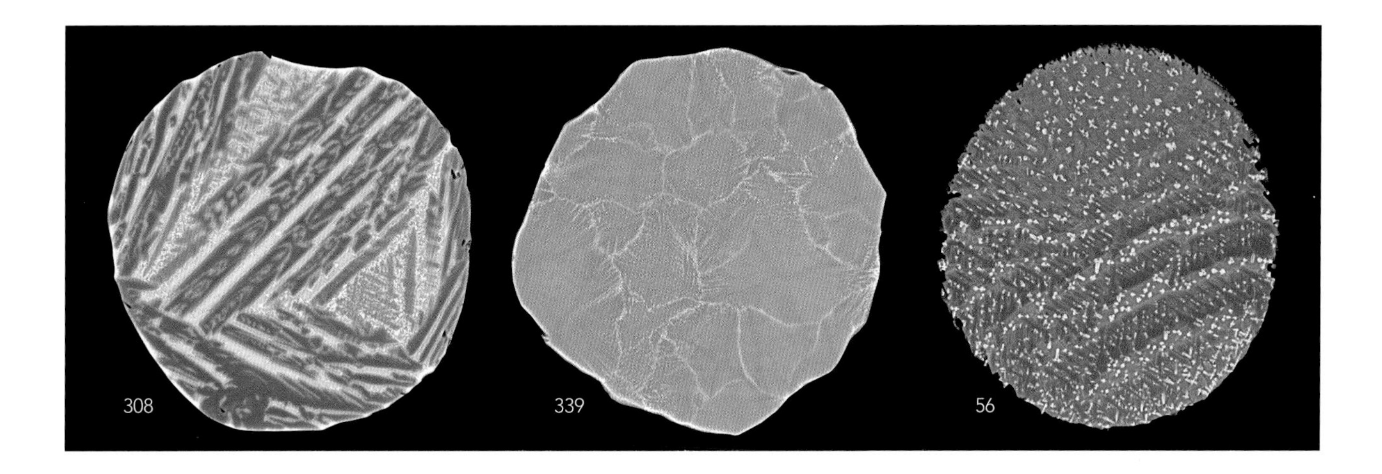

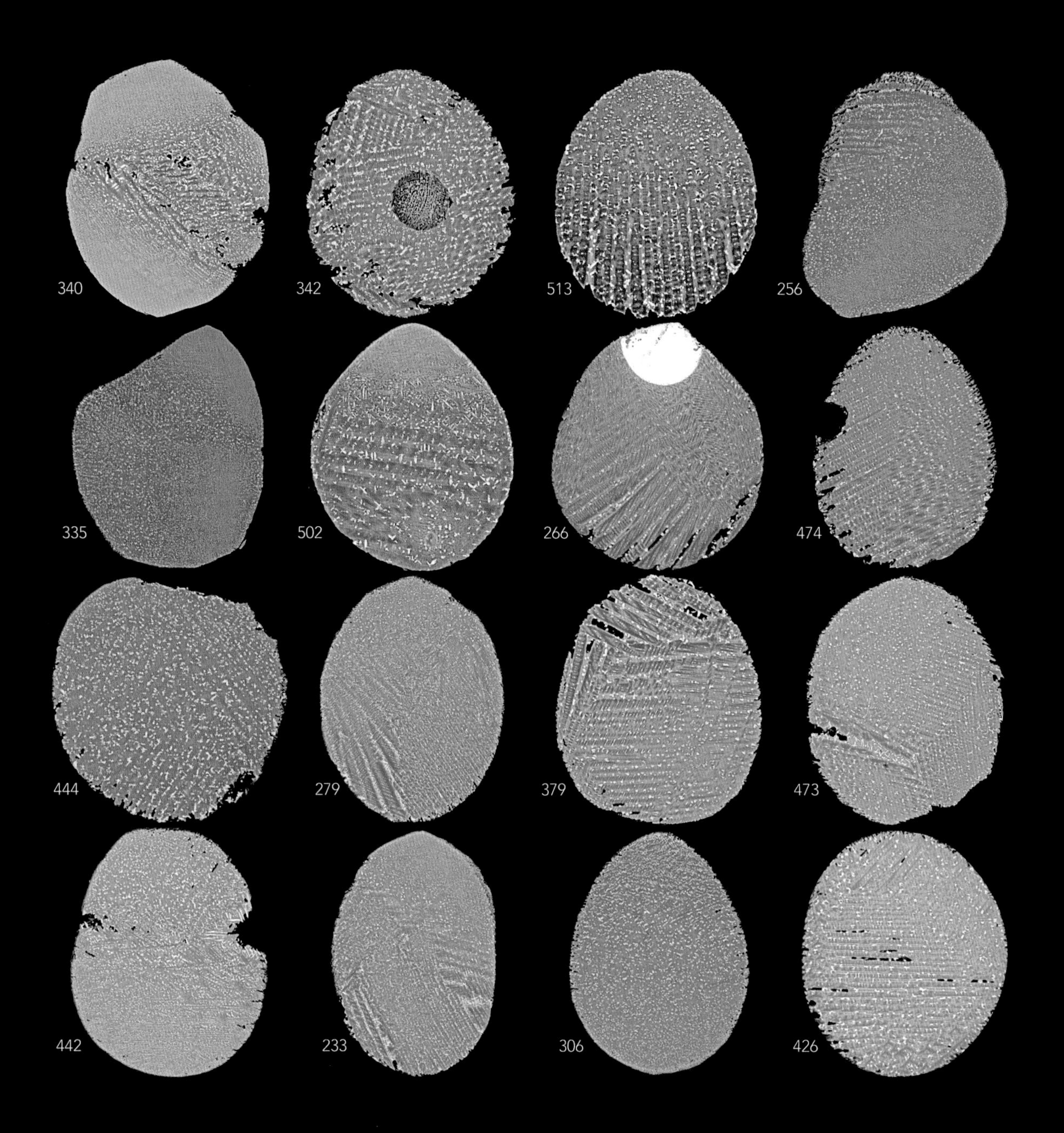
340
342
513
256
335
502
266
474
444
279
379
473
442
233
306
426

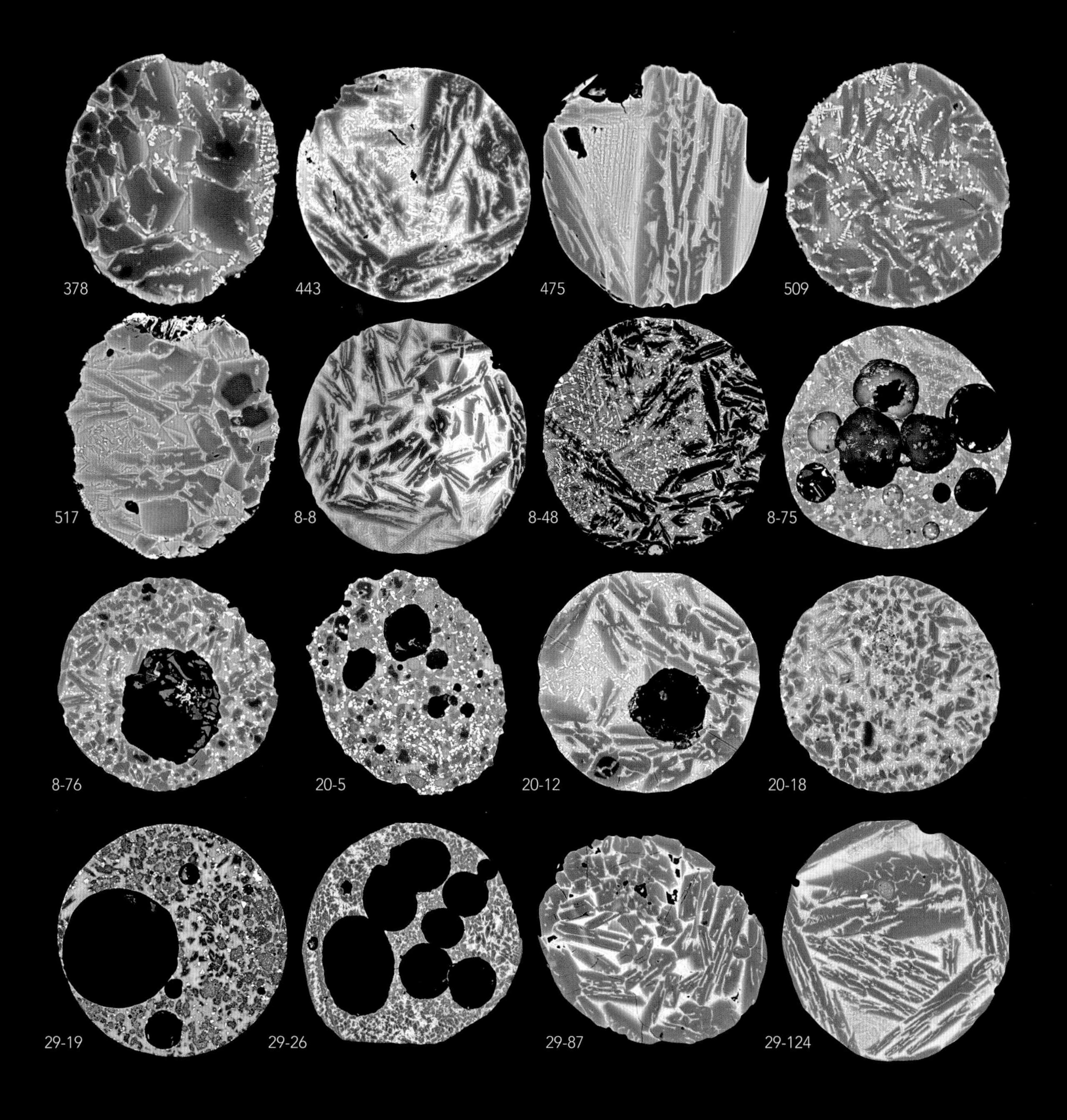
378
443
475
509
517
8-8
8-48
8-75
8-76
20-5
20-12
20-18
29-19
29-26
29-87
29-124

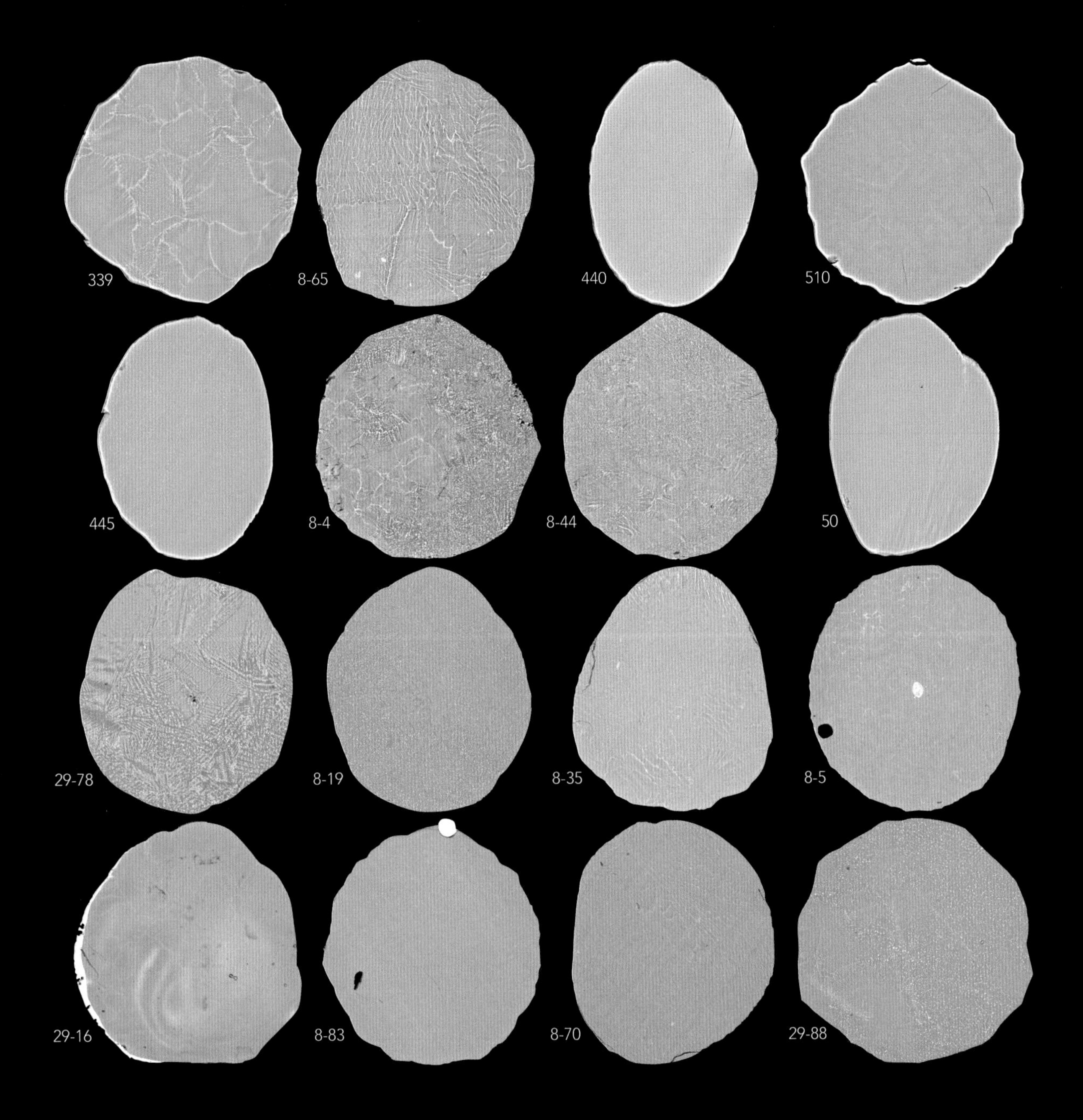
339
8-65
440
510
445
8-4
8-44
50
29-78
8-19
8-35
8-5
29-16
8-83
8-70
29-88

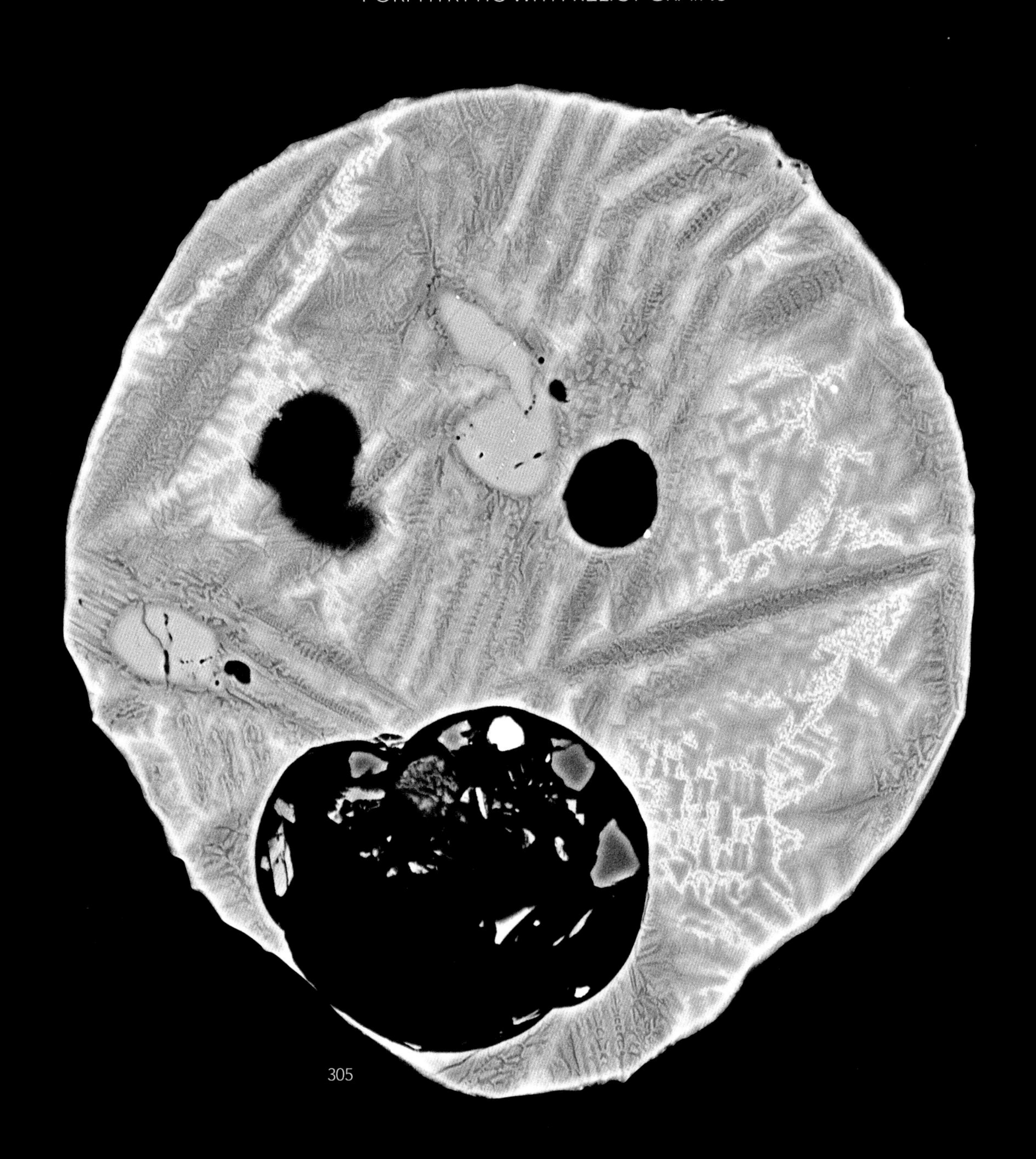

305

GLASS MICROMETEORITES

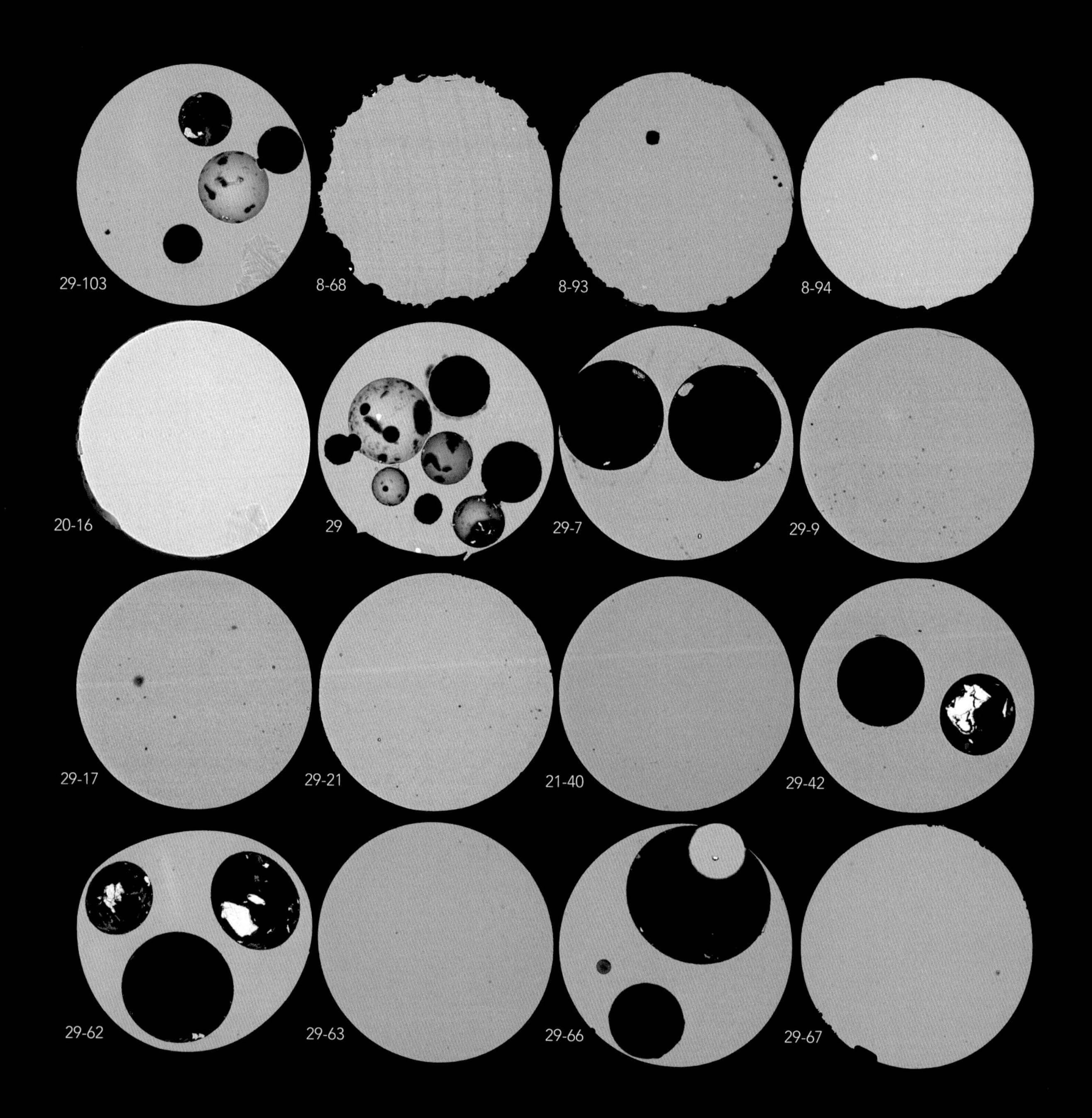

VARIOUS MICROMETEORITES

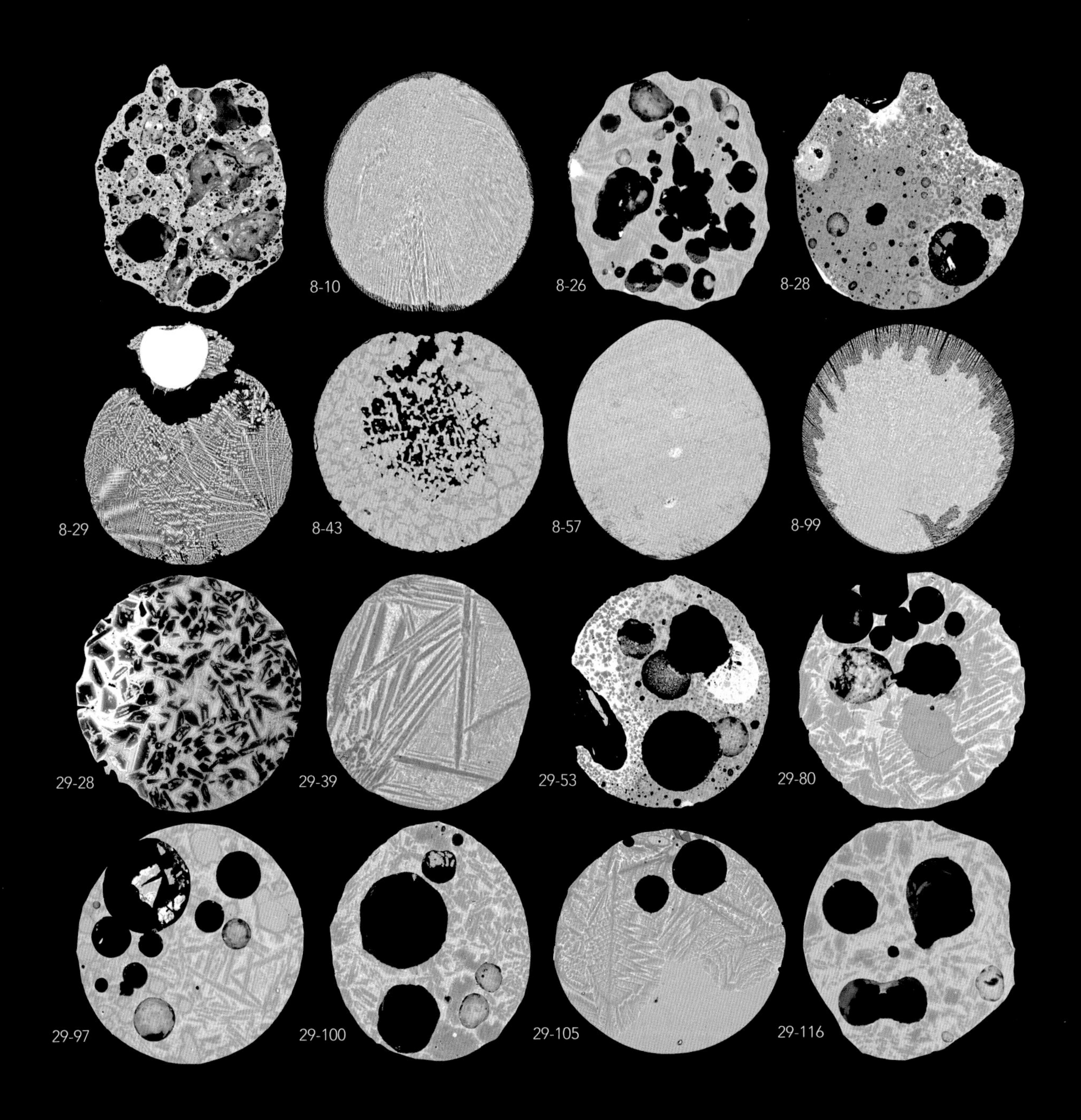

The New Collection

Previous studies have primarily concentrated on the various aspects of the chemistry, formation, verification and classification of micrometeorites. For this purpose backscatter SEM section images (pages 16–23) are usually the most relevant. In order to search for micrometeorites in populated areas, however, it is imperative to know what they look like on the outside—their morphology.

The following part of the book presents an overview in photos and scanning-electron microscope (SEM) images of micrometeorites from the new collection. This is the first time the morphology of micrometeorites is systematically documented in high-resolution color photography. The new micro photo techniques were developed together with Jan Braly Kihle especially for this project.

For many years meteorite hunters have built micrometeorite traps of various types. Some have succeeded, like the water traps to catch unmelted micrometeorites, but given the low influx rate a really efficient trap would have to be much larger. In order to catch thousands of cosmic spherules the trap would have to be the size of a football field (or larger), and accumulate particles over decades. The challenges connected with the construction of something like that have discouraged more than one good scientist. There are however such areas already in place, possibly in your neighborhood, and ripe for harvesting: roofs.

The micrometeorites in the new collection are mainly found on the roofs of buildings with a maximum of 50 years of age, so it can be assumed that the stones have a terrestrial age of 0–50 years, which make them fresh compared with most of the micrometeorites in the other collections. Most of the Antarctic stones, for example, have a terrestrial age of 1,000–700,000 years (except the Concordia collection, from melted snow), and are weathered accordingly.

By monitoring a skyward facing area like a roof at regular time intervals, it should be possible to be even more precise in future sampling, perhaps down to the week (or even day) the micrometeorite fell to Earth. With careful preparation (cleaning of the collecting area) around the annually reoccurring meteor showers it should be possible to identify material from some of the comets, possibly also to detect variations in the influx rate over time.

Each of the micrometeorites on the following pages are marked with its catalog number referring to the NMM (Norwegian micrometeorite) database, with the complete history of each stone—when/where/how it was found, which analysis it has undergone, a rough classification, etc. These data are not included in the book, but can be obtained from the author upon request. Furthermore, in the pictures I have not included the size of each stone, as the variations are small. Practically all are between 0.2–0.3 mm, with a few exceptions, like the 0.5 mm giant on the next page. All the stones were found at various locations in Norway during 85 field searches in 2015, with one exception: micrometeorite number 199 (see page 53) was found at Fontainebleau, France.

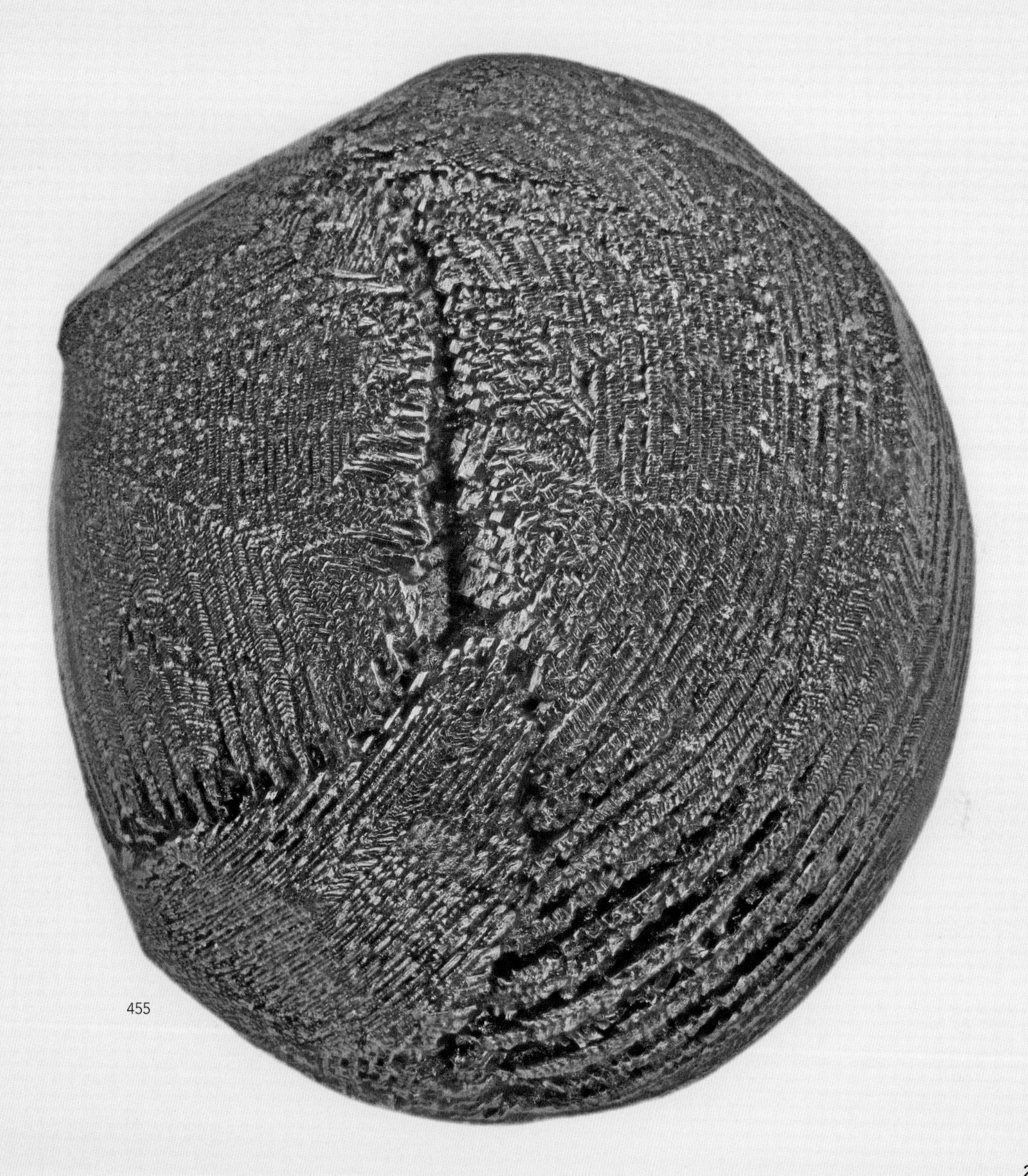

455

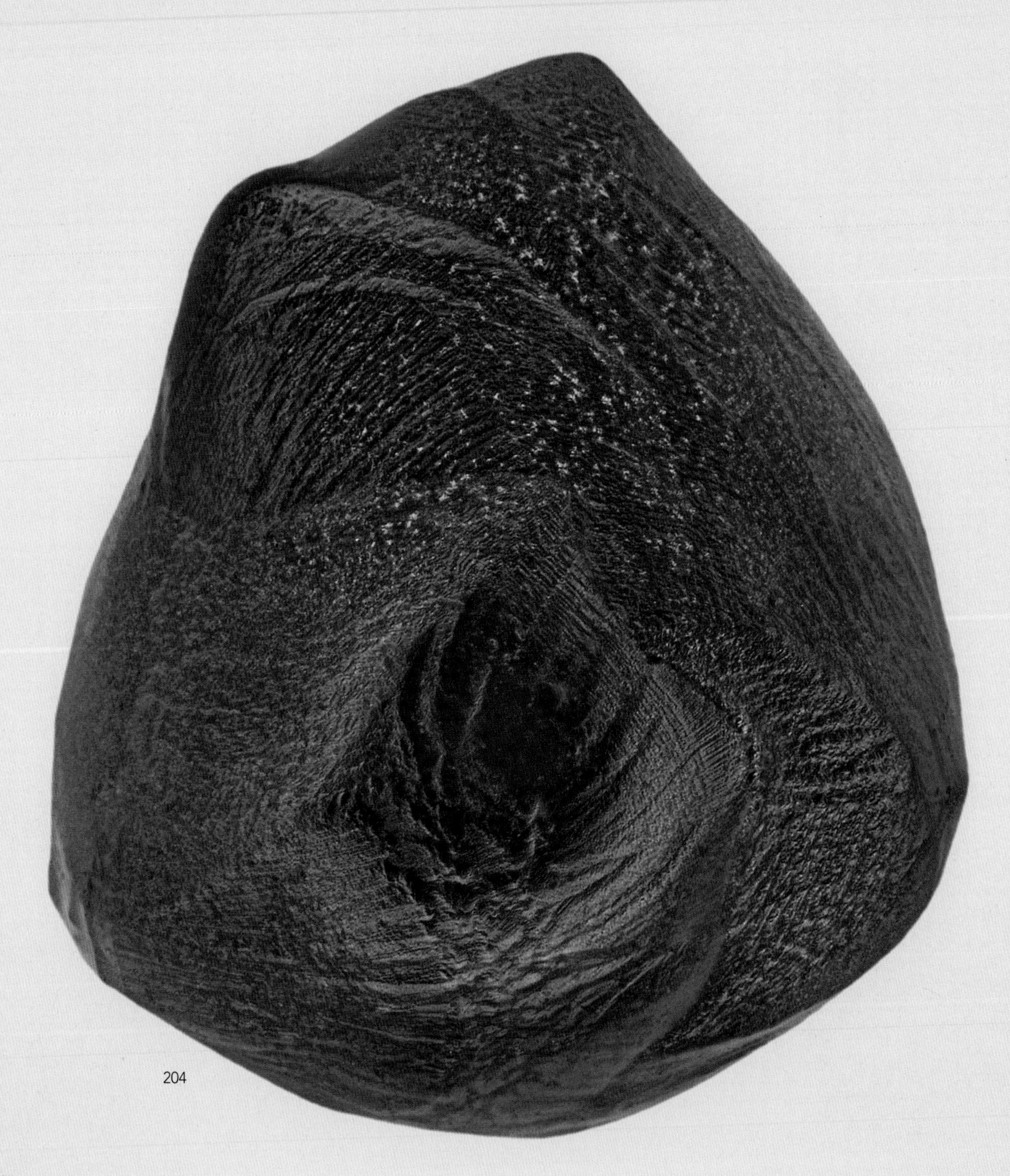

204

CRYPTOCRYSTALLINE MICROMETEORITES

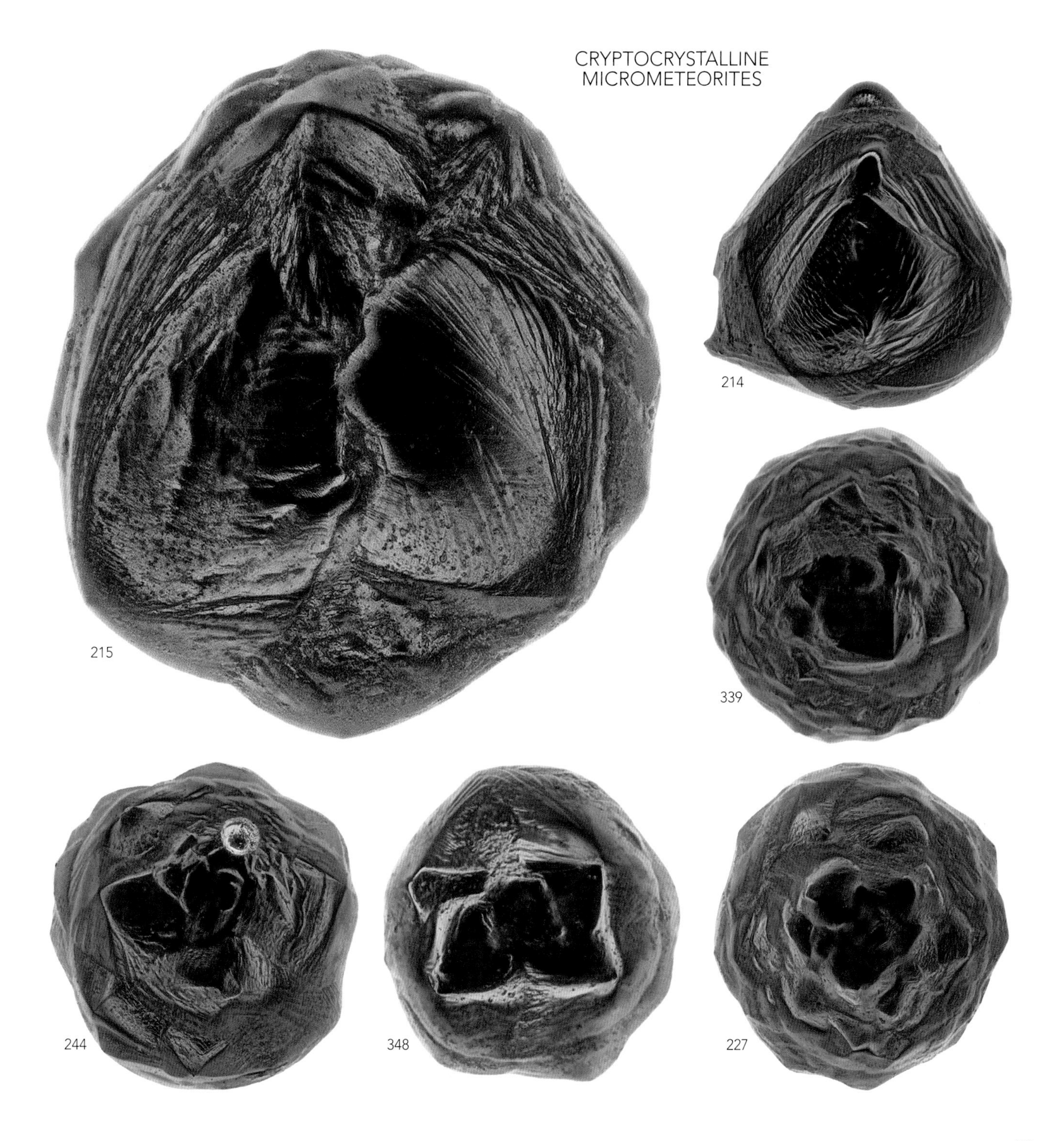

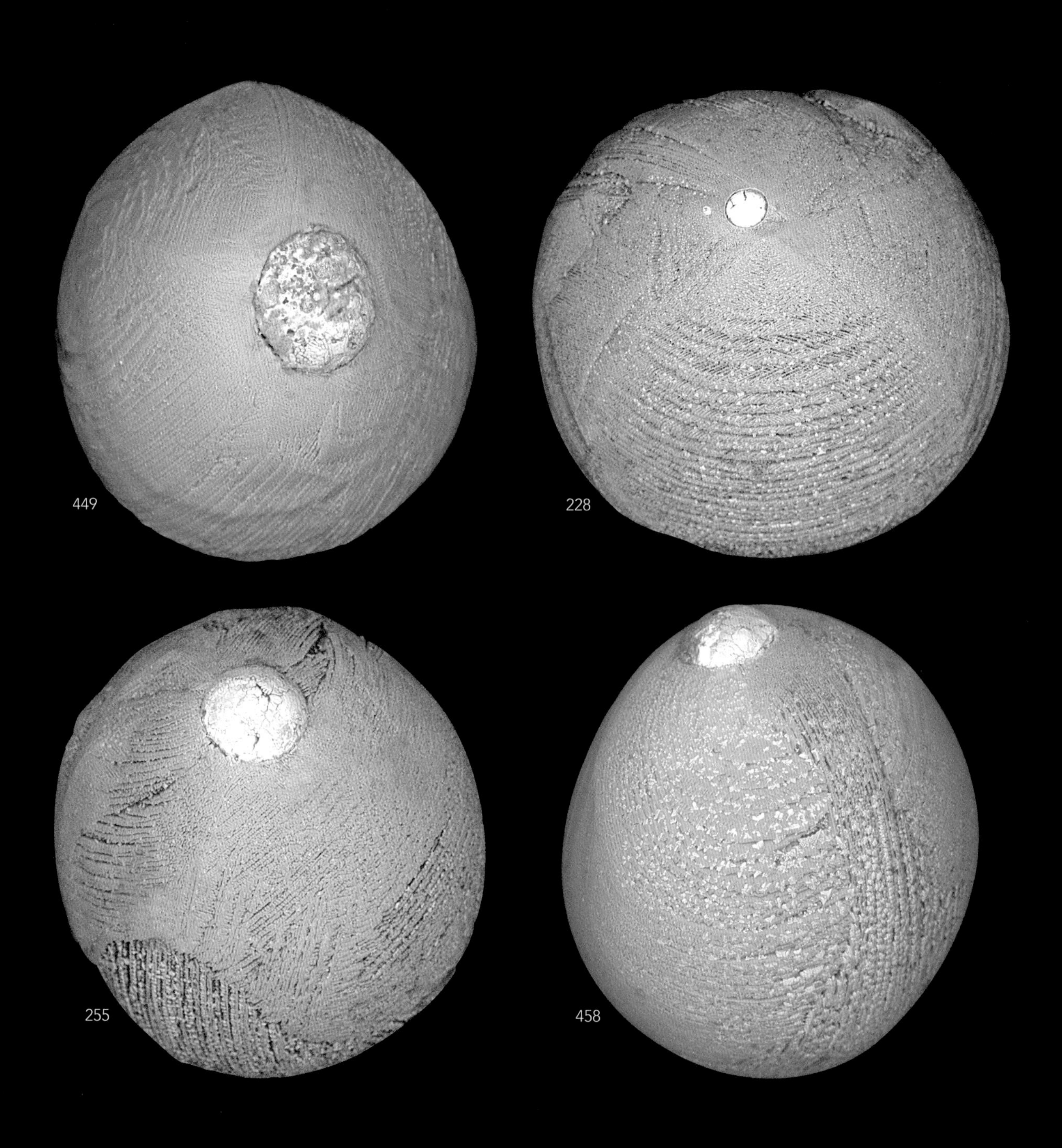
449
228
255
458

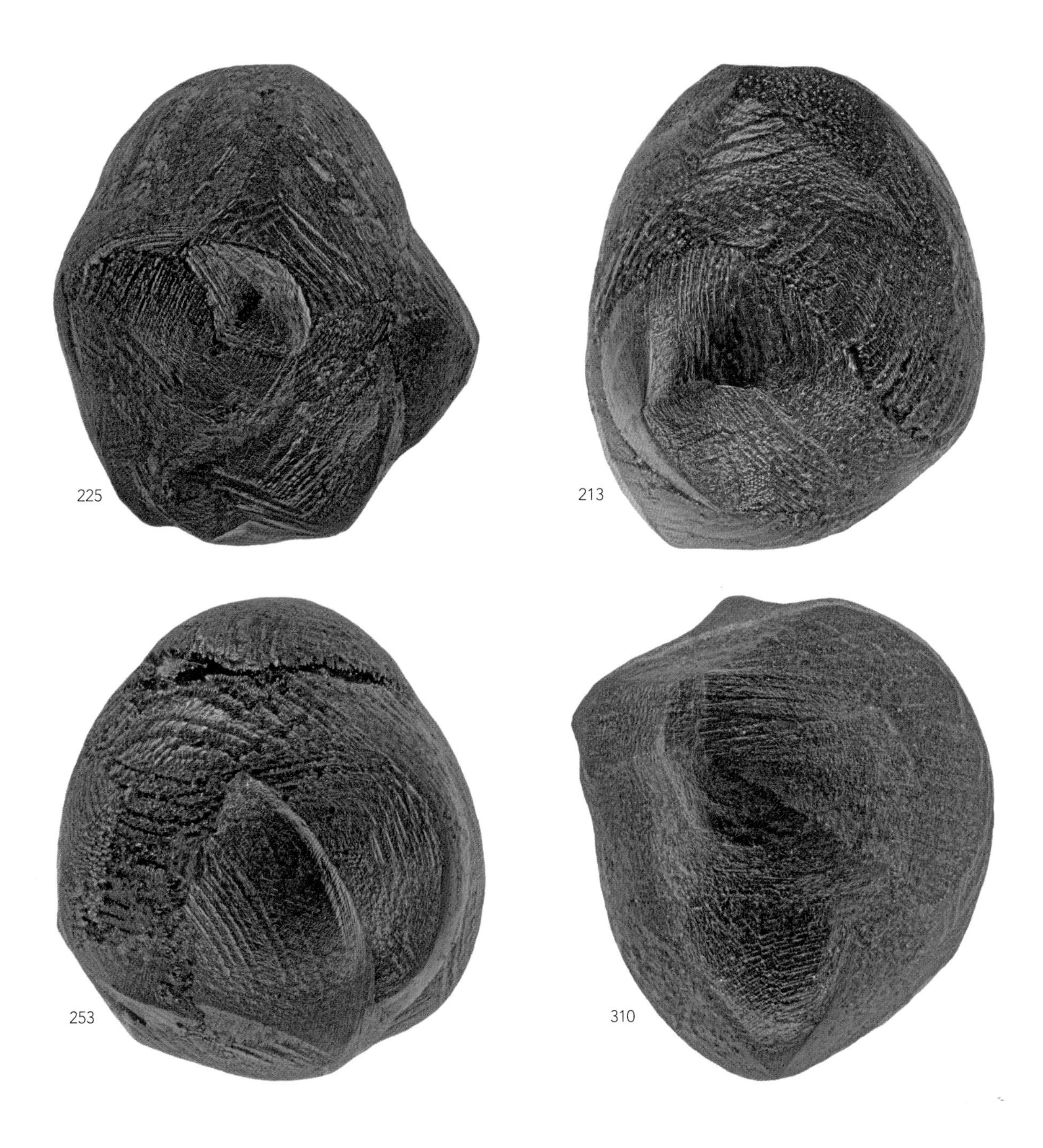
225
213
253
310

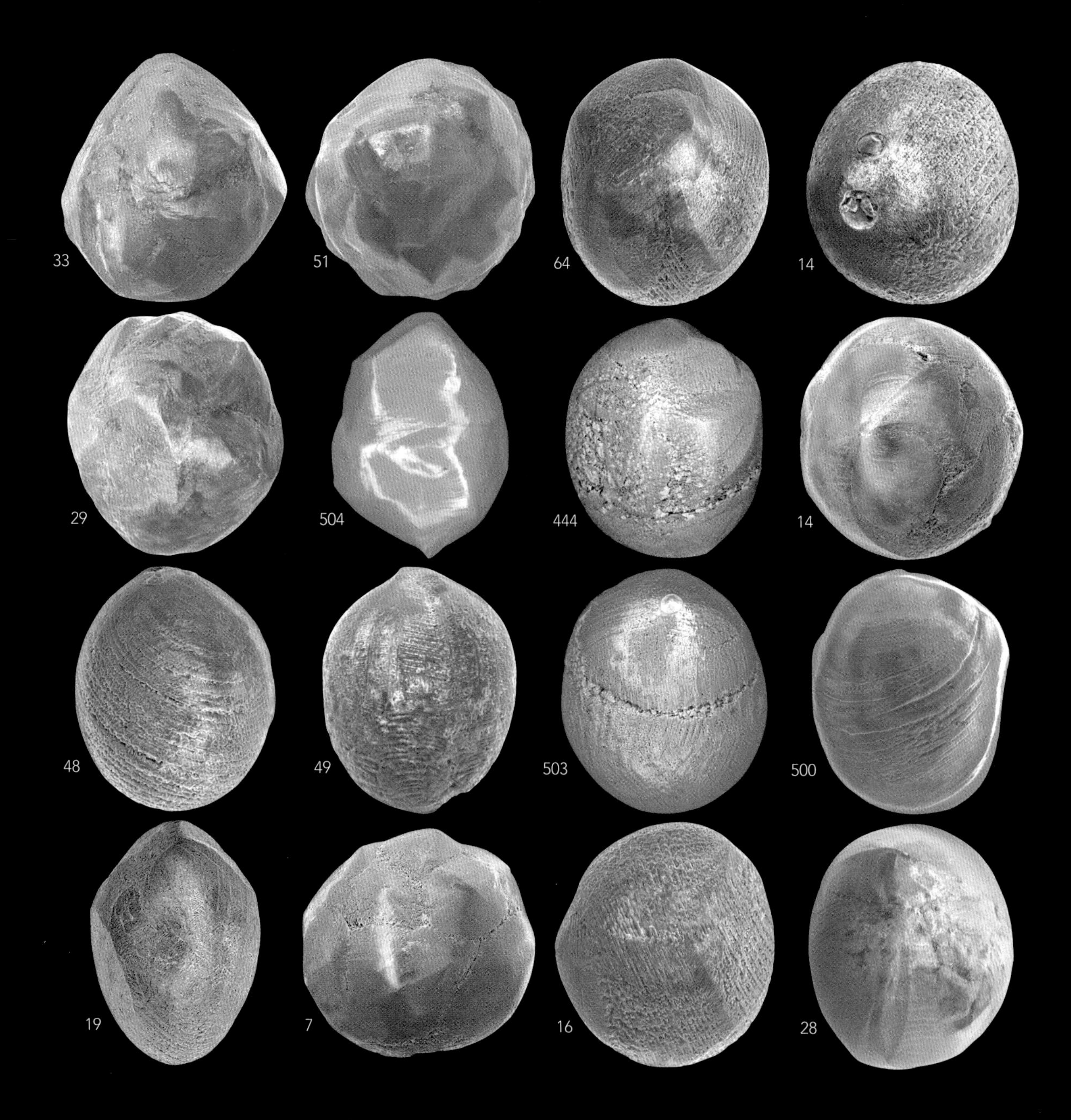
33
51
64
14
29
504
444
14
48
49
503
500
19
7
16
28

338

6

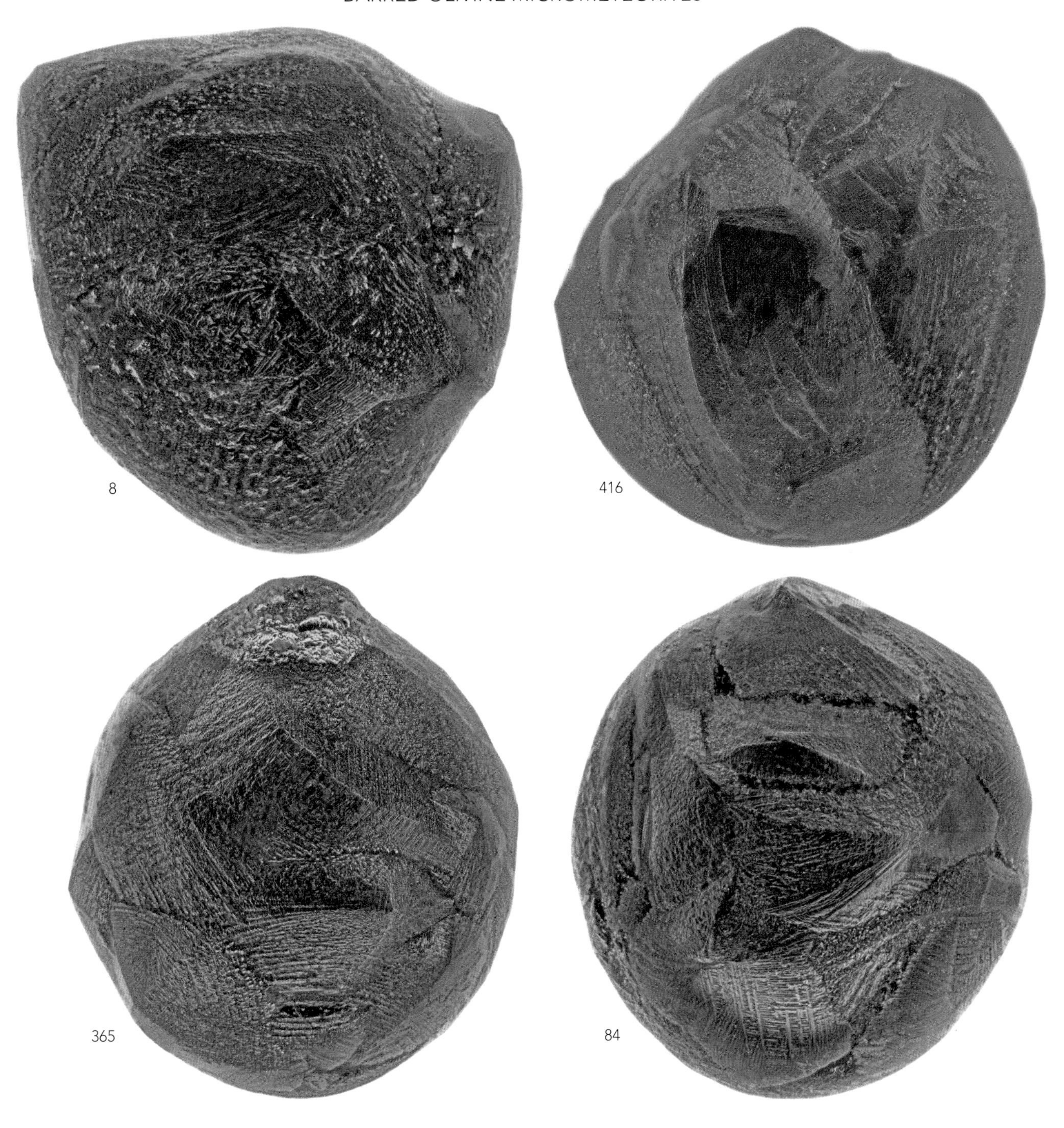
8
416
365
84

30

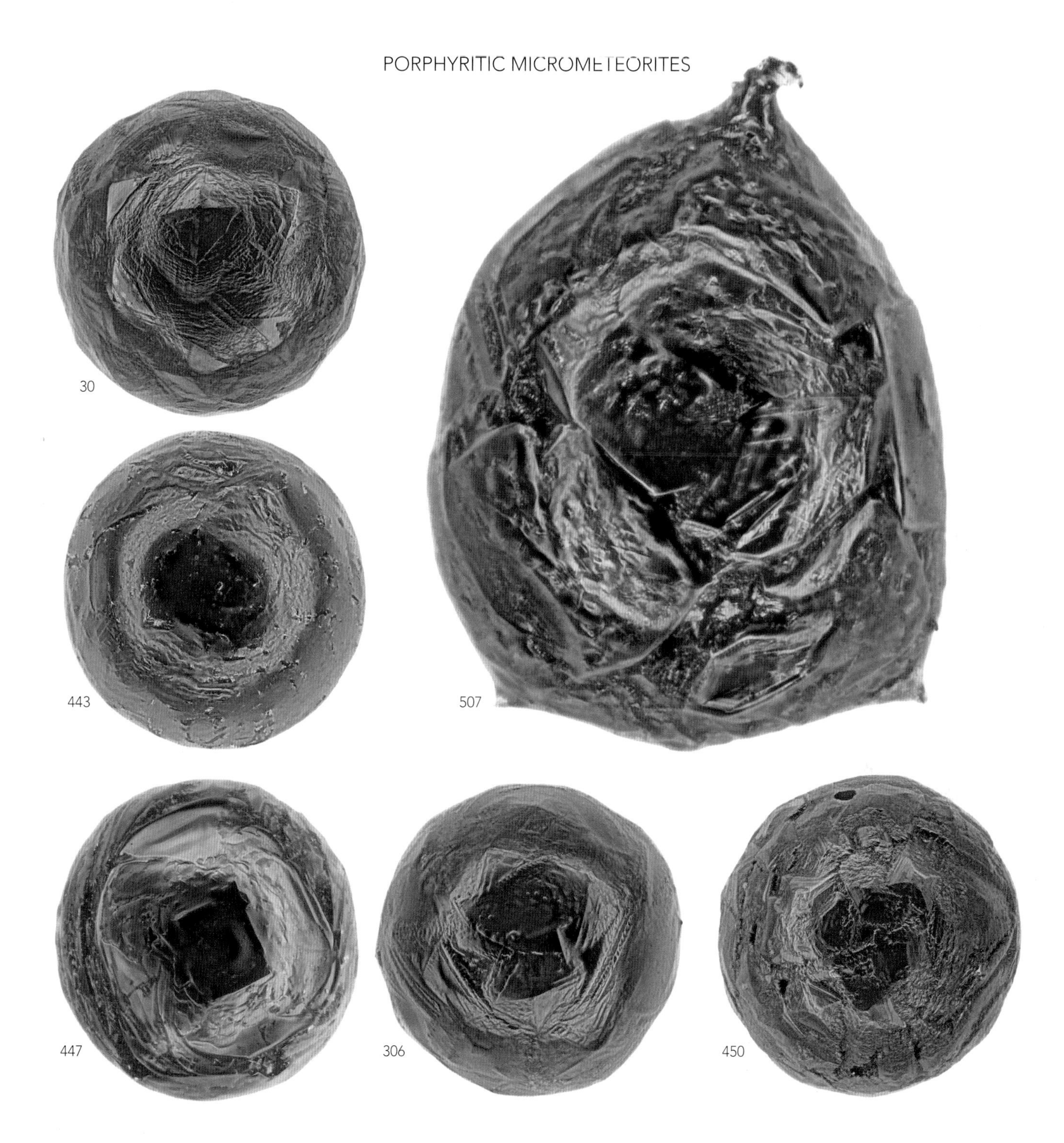
30
443
507
447
306
450

447

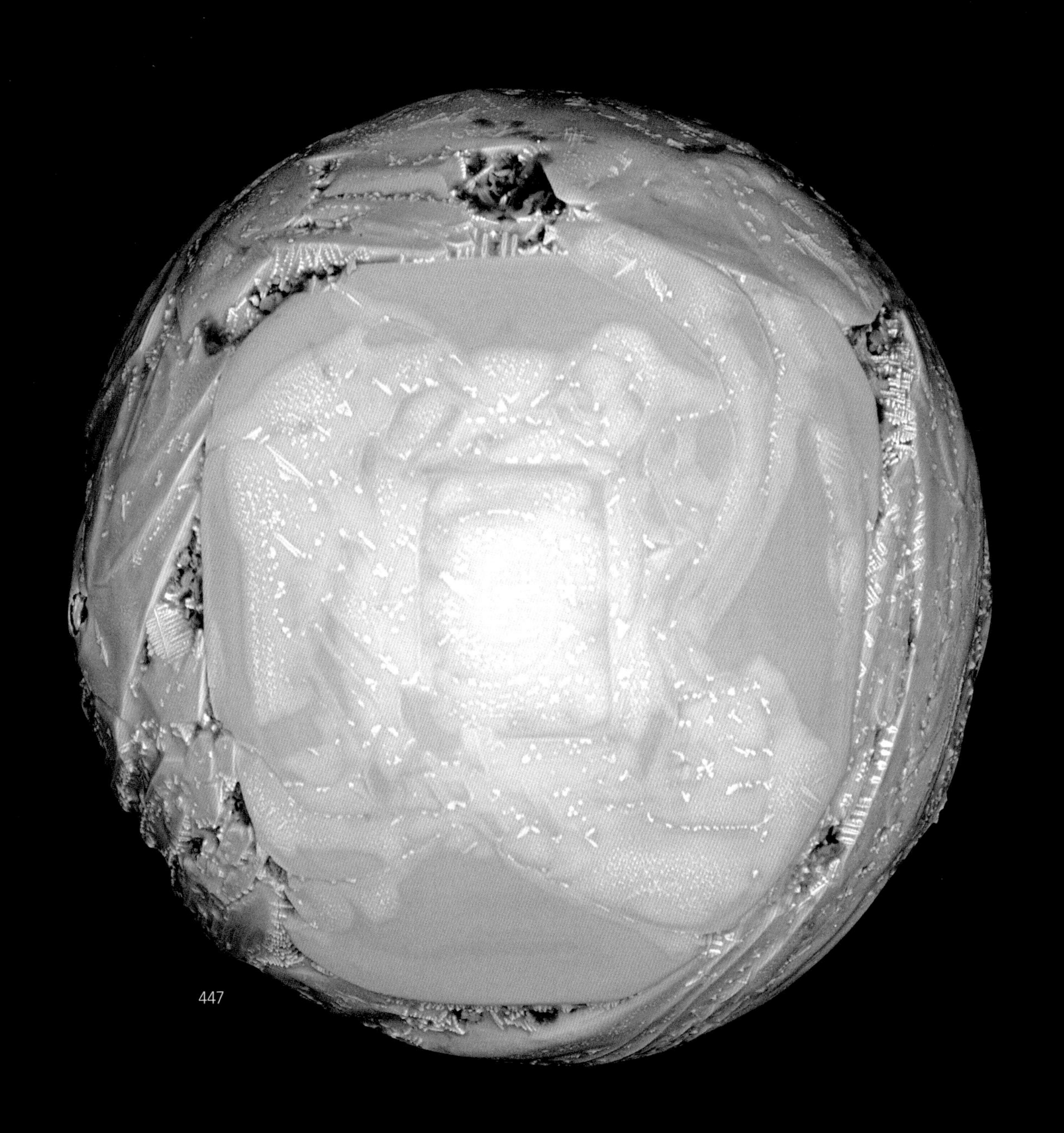

447

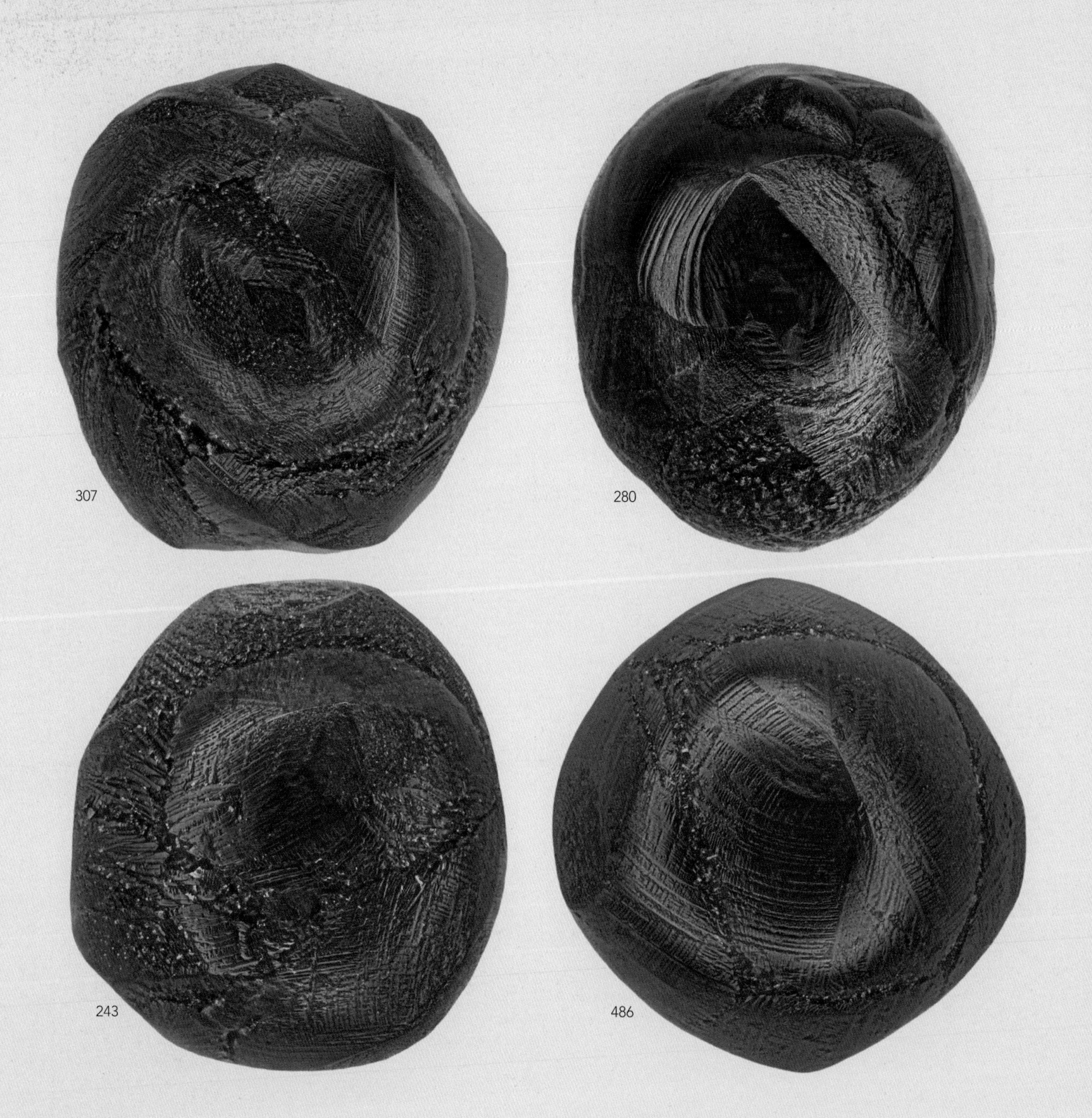
307
280
243
486

340

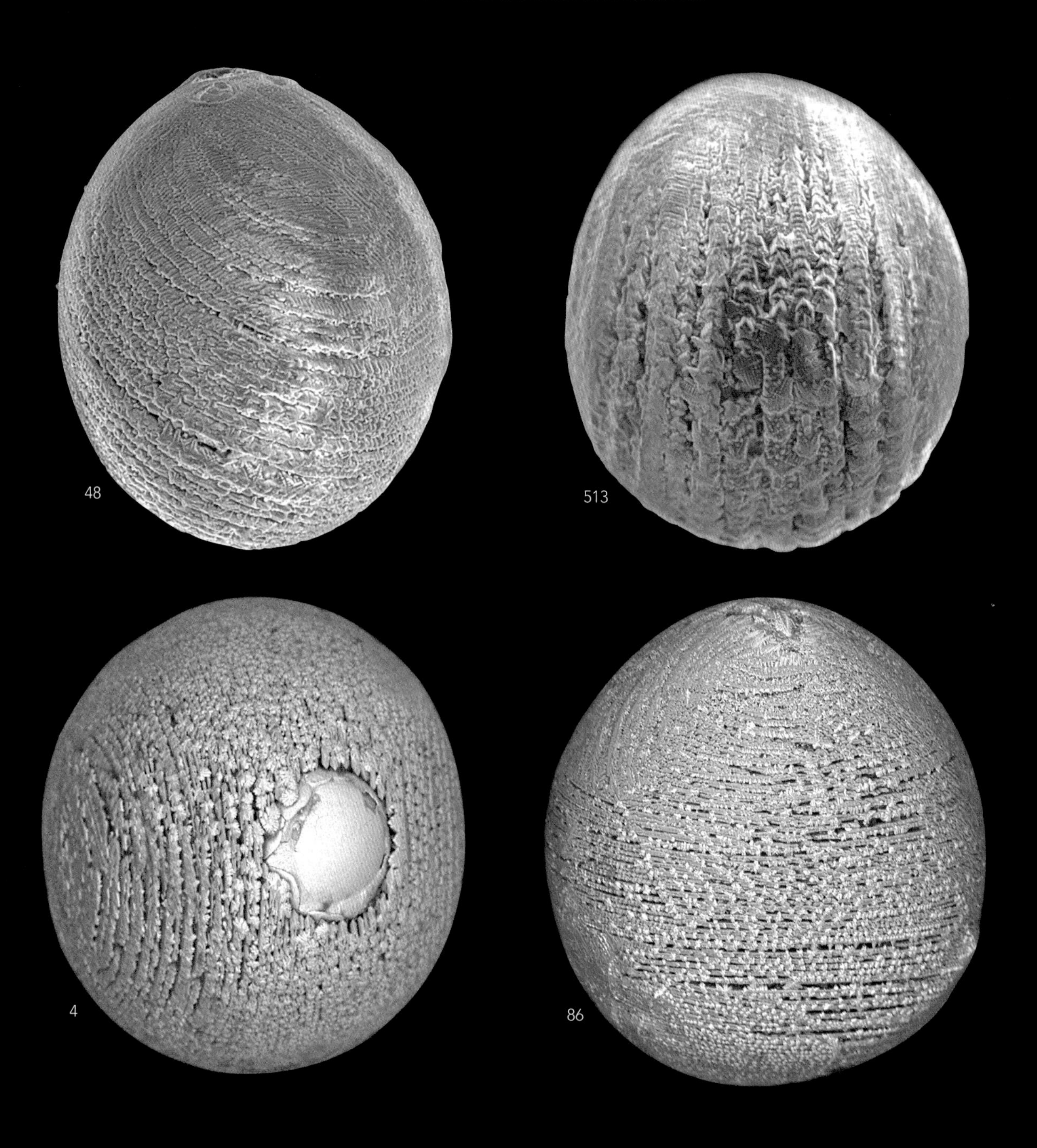
48
513
4
86

226

417
341
337
478
476
421

500

516

11
436
337
478
341
421

10

445
441
476
366
279
74

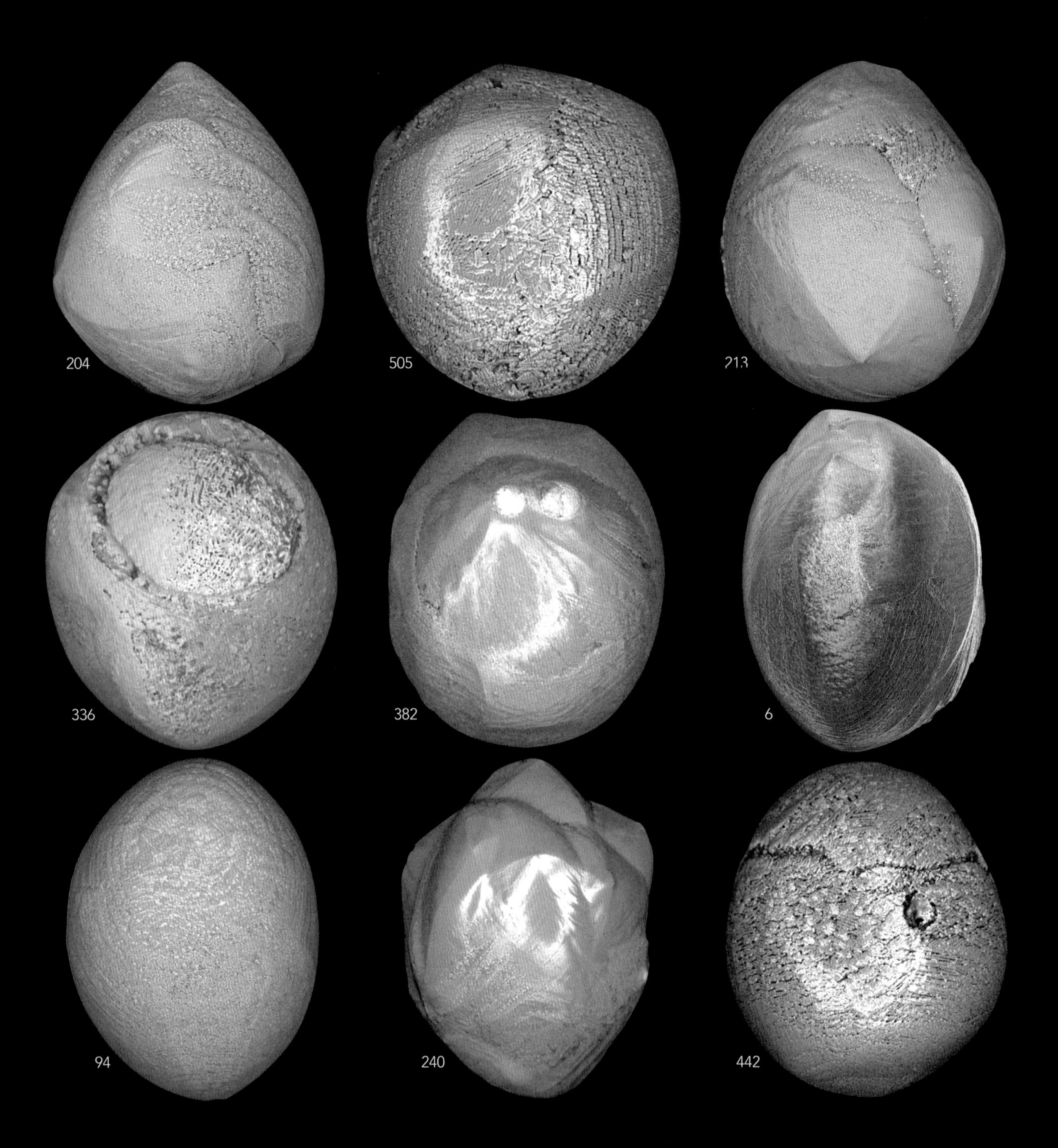
204
505
213
336
382
6
94
240
442

73 440 349

421 375 380

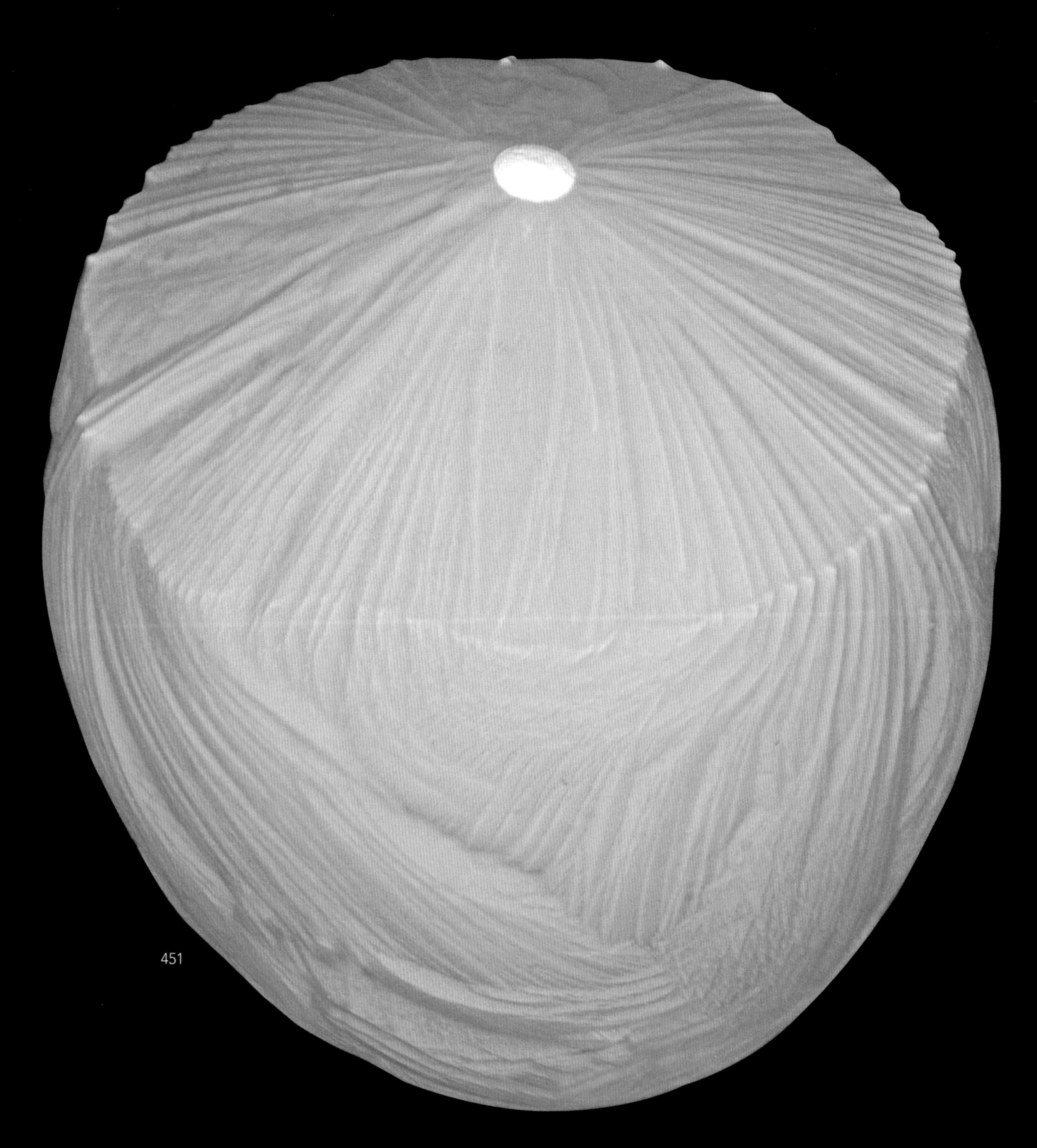

451

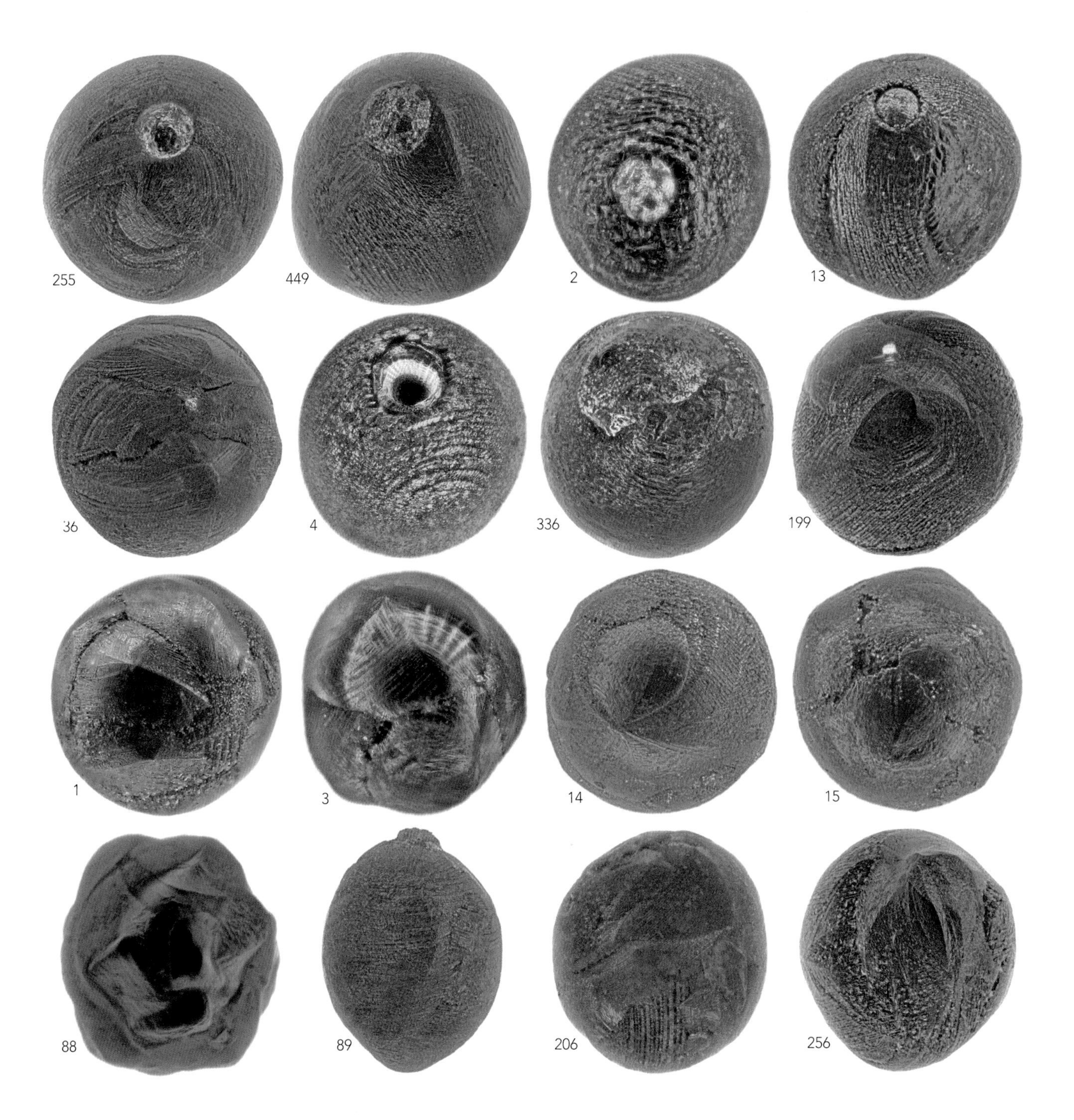
255
449
2
13
36
4
336
199
1
3
14
15
88
89
206
256

458

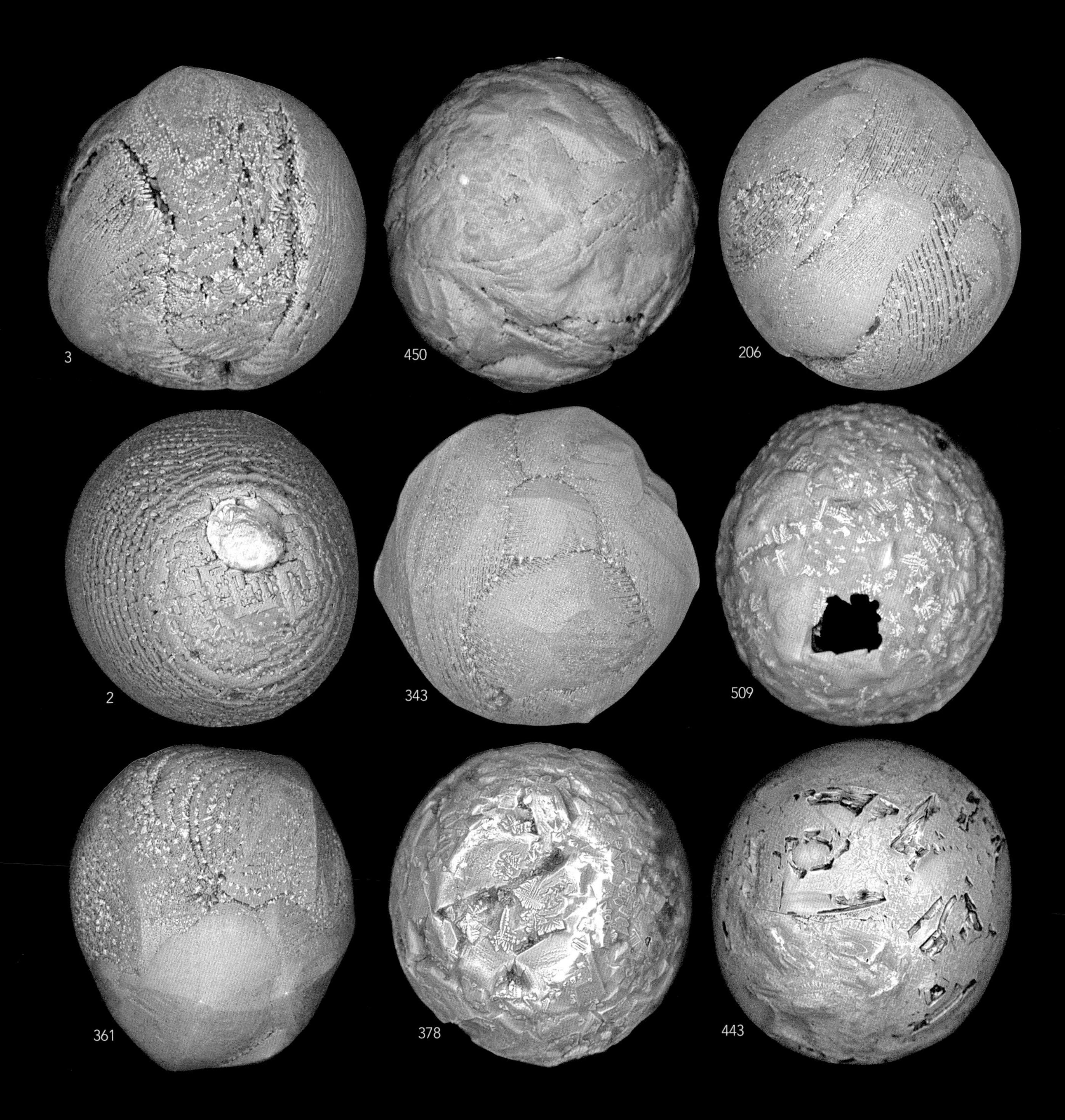
3
450
206
2
343
509
361
378
443

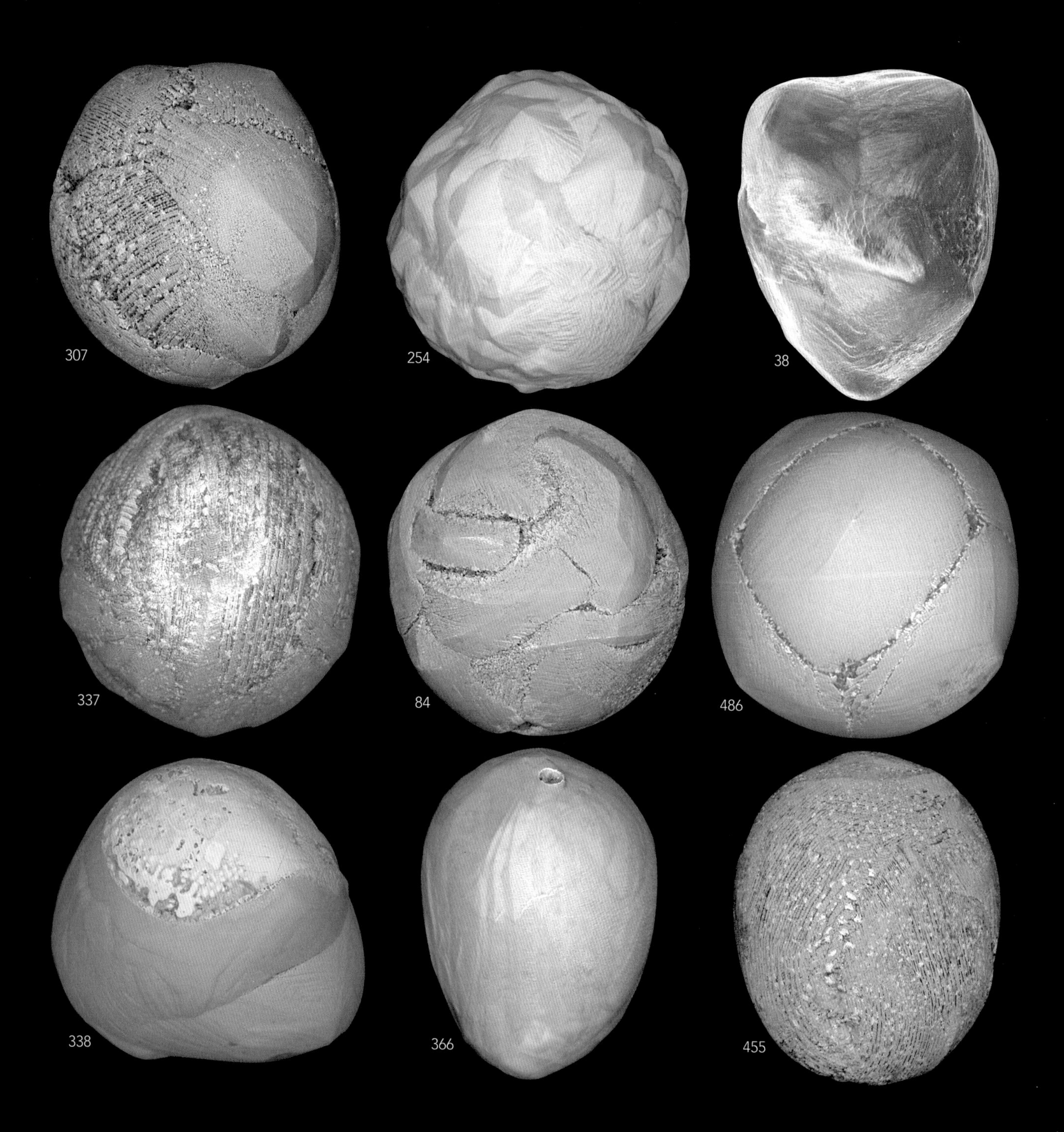
307
254
38
337
84
486
338
366
455

362

59

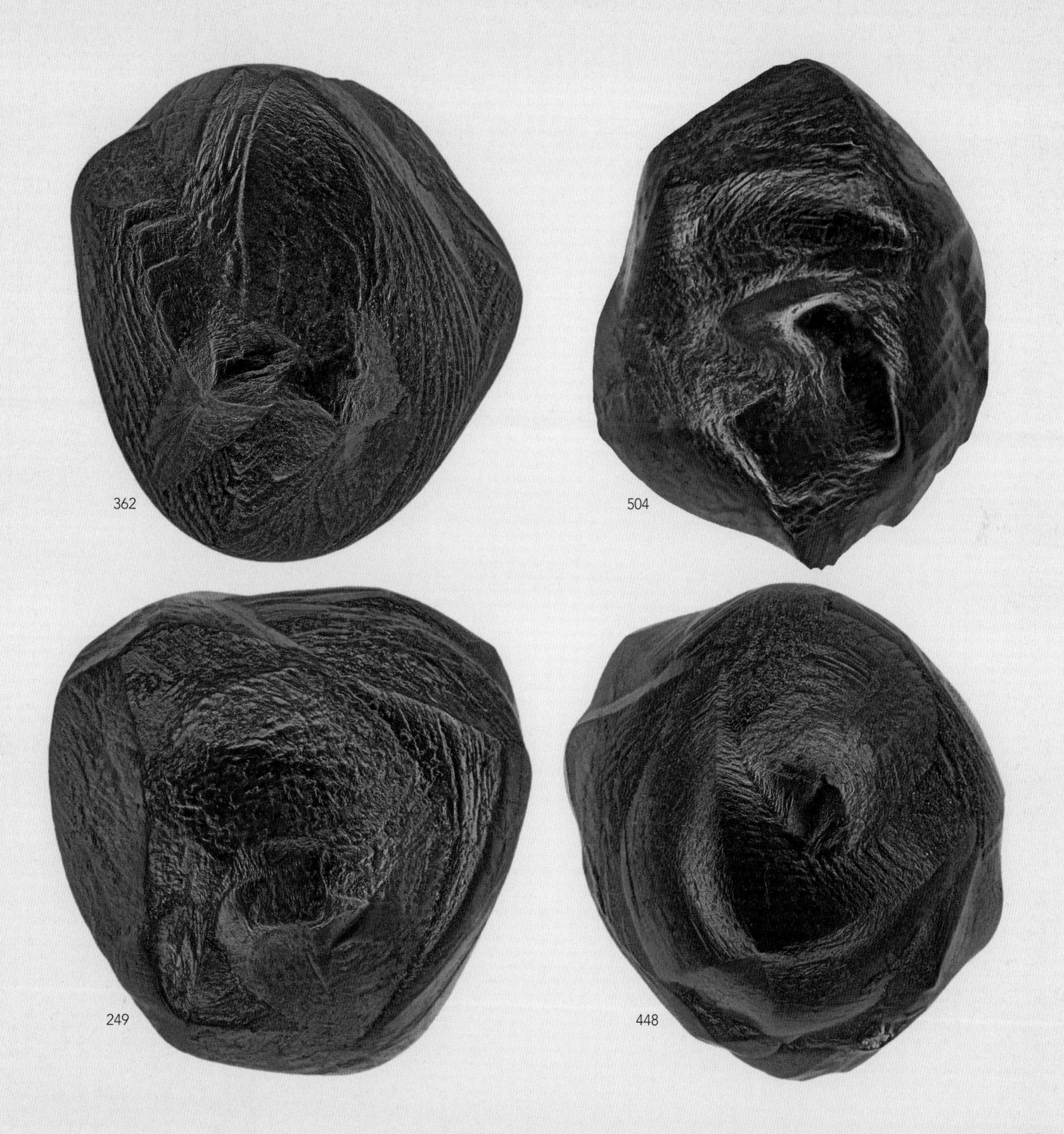
362
504
249
448

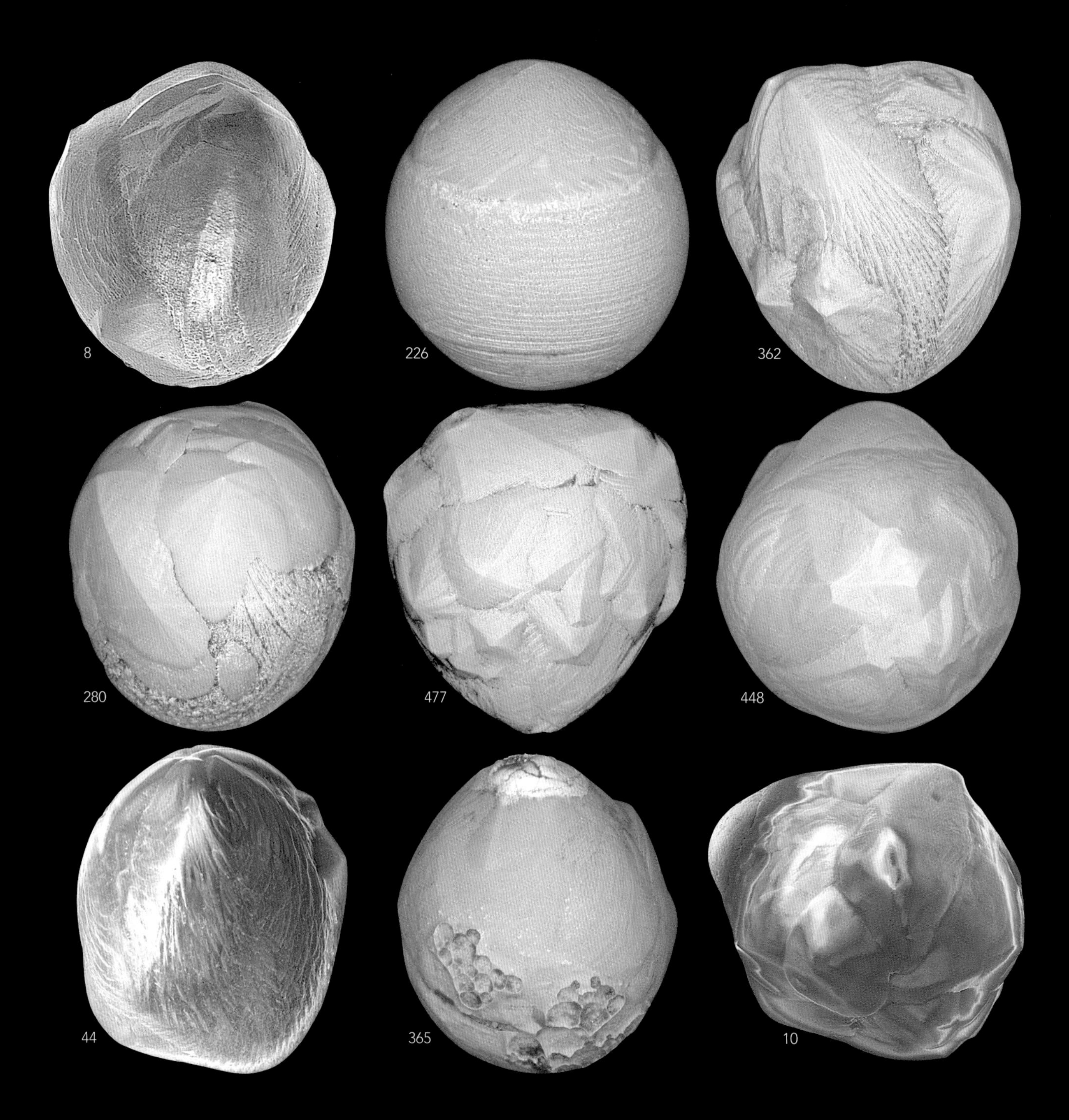
8
226
362
280
477
448
44
365
10

451

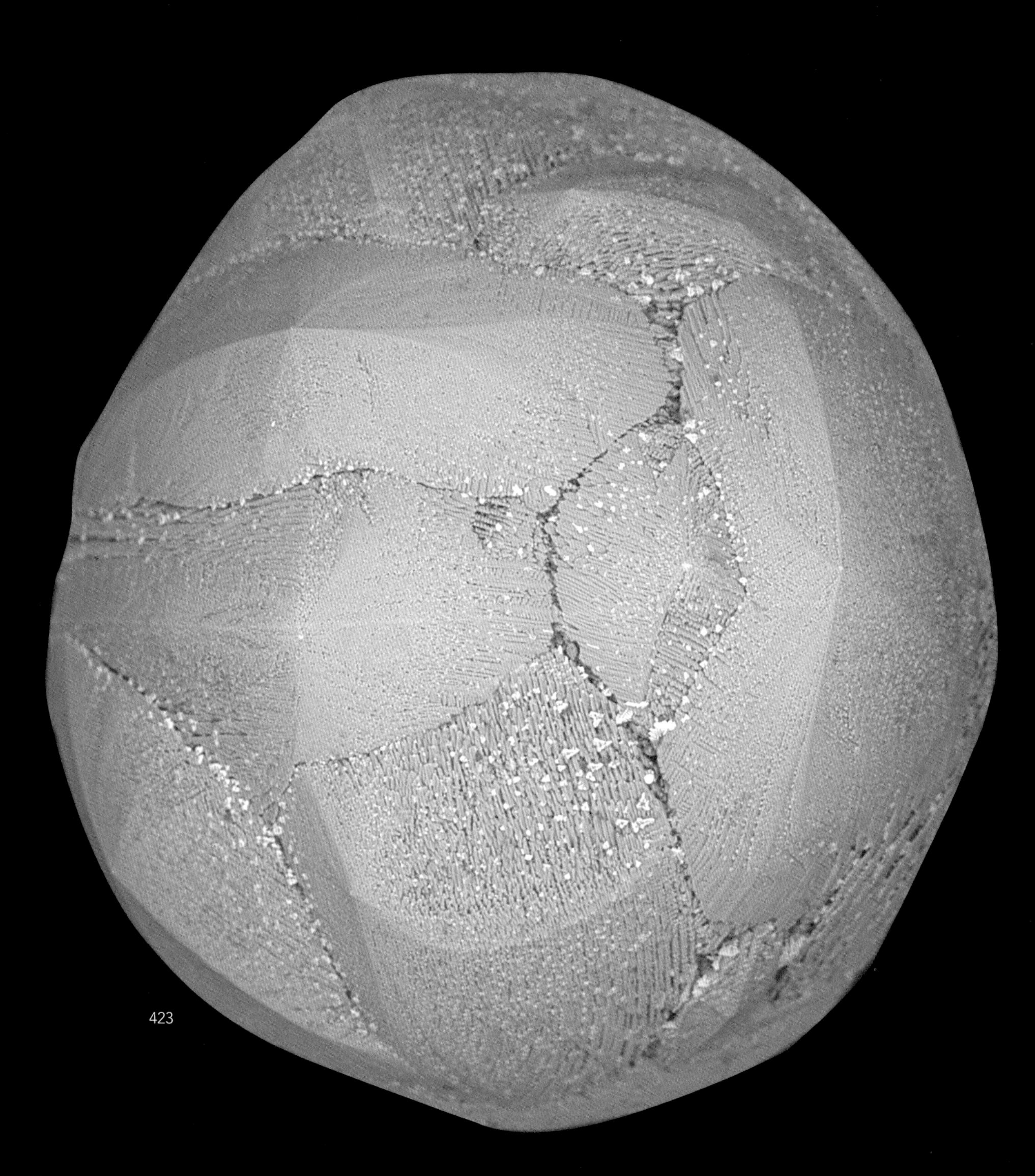

423

443

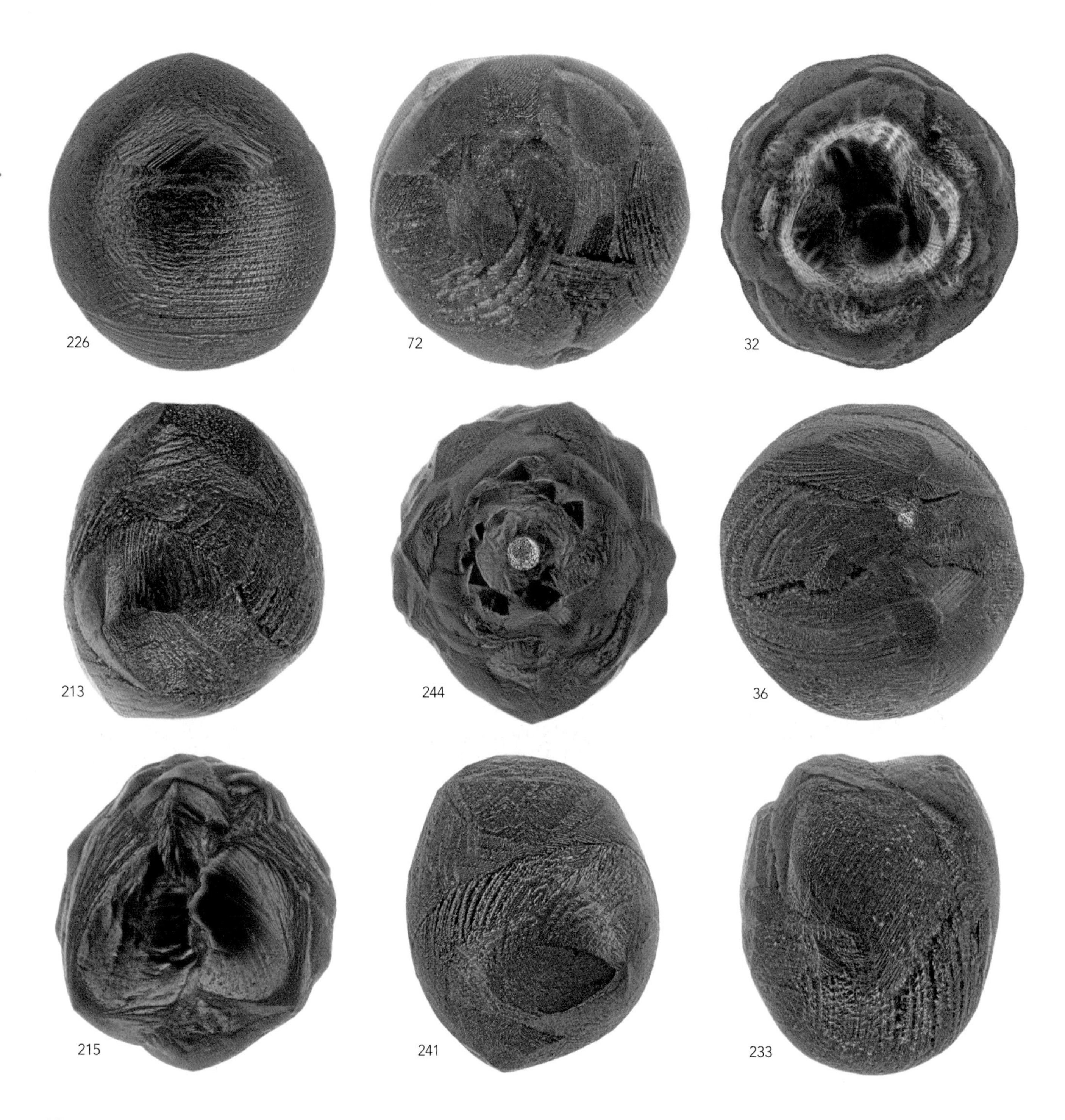
226
72
32
213
244
36
215
241
233

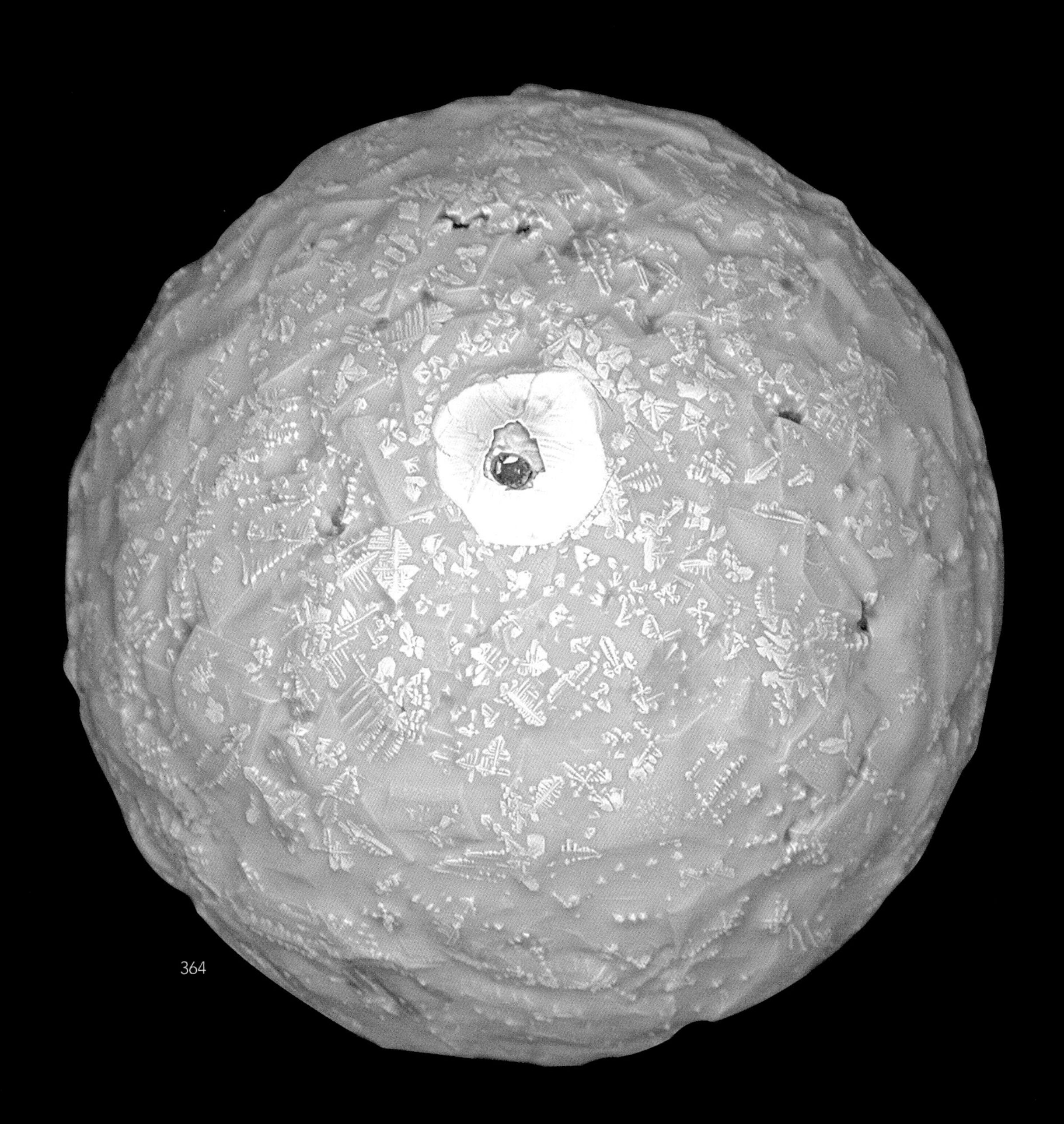

364

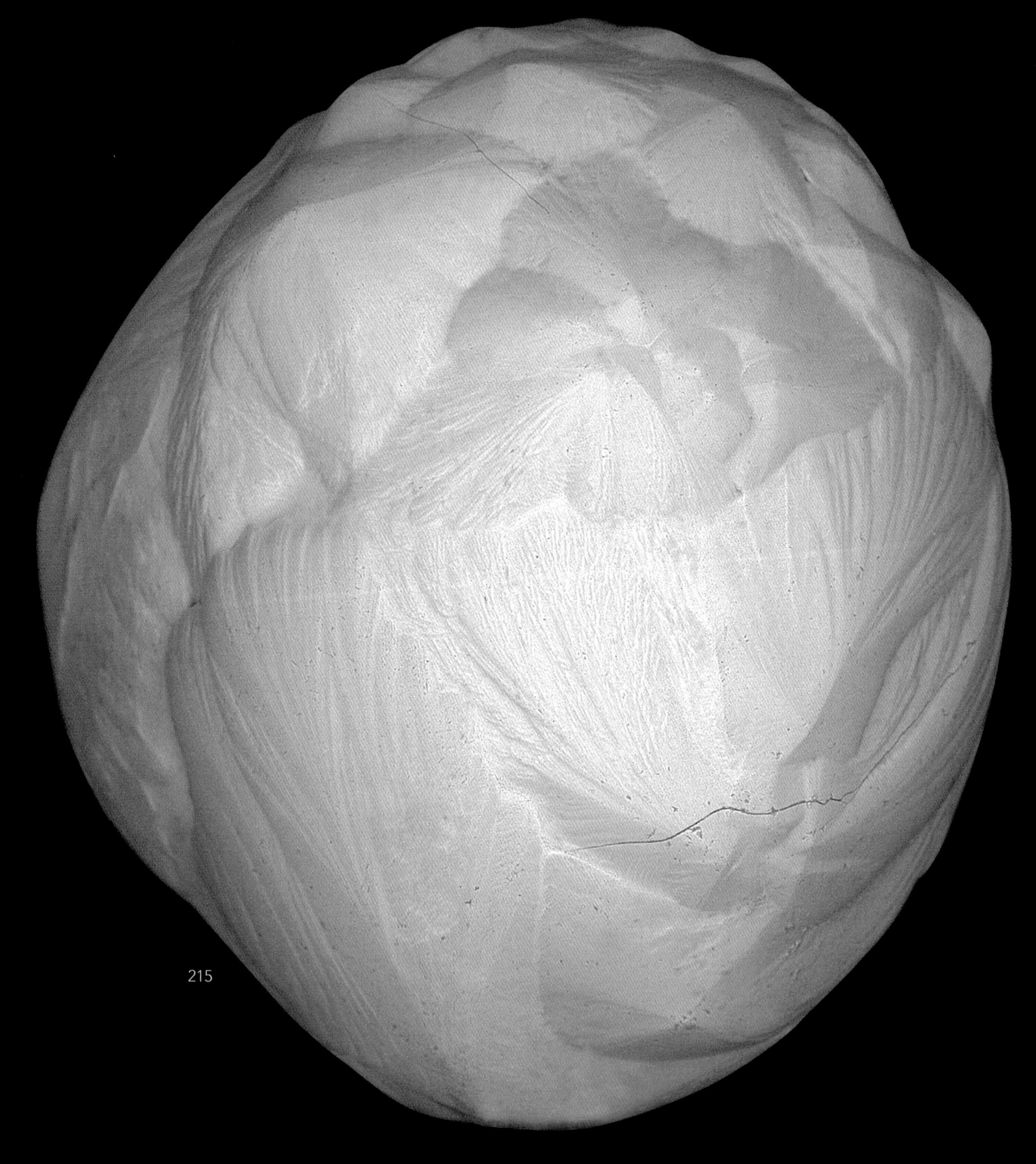

215

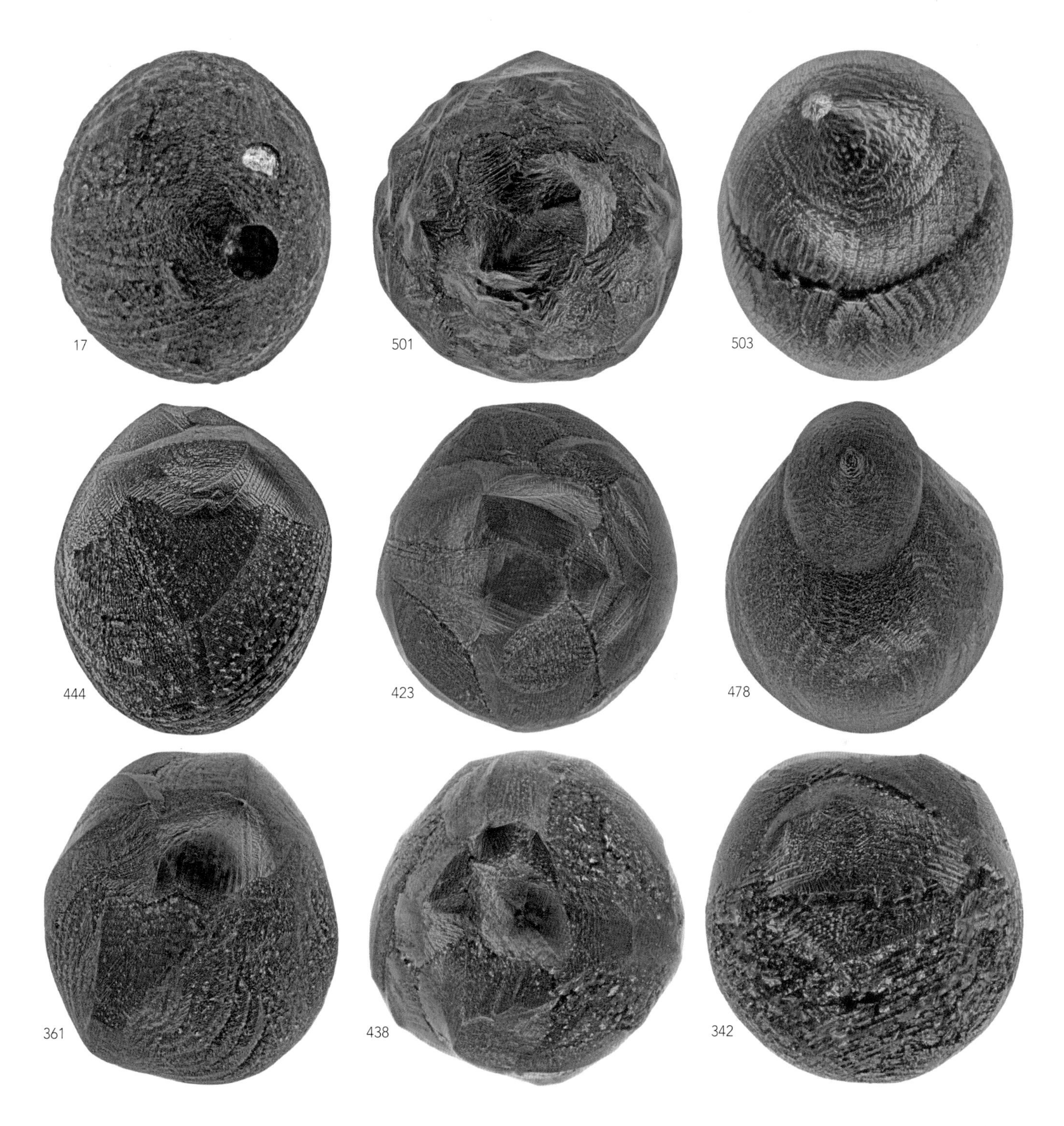
17
501
503
444
423
478
361
438
342

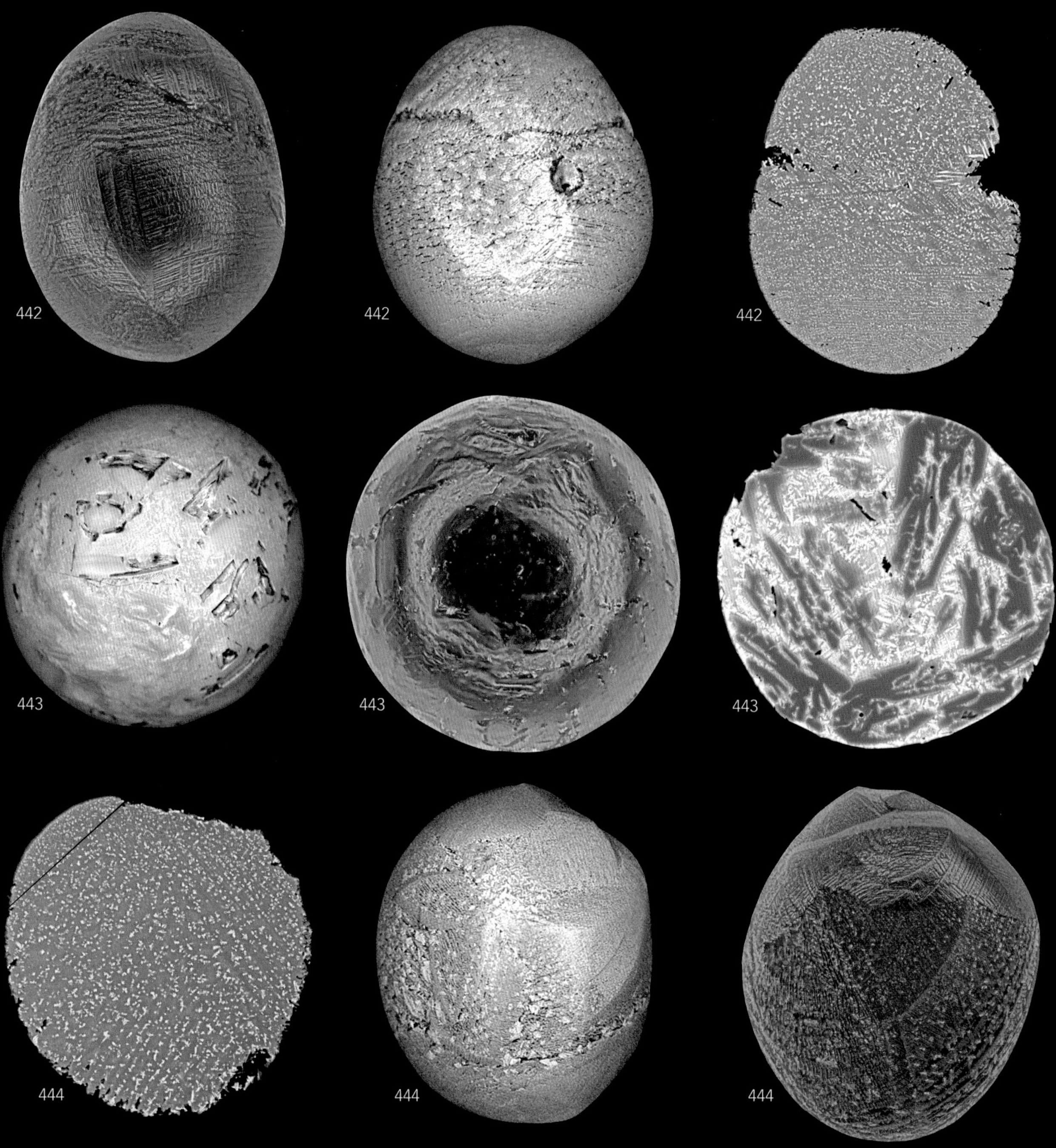
442
442
442
443
443
443
444
444
444

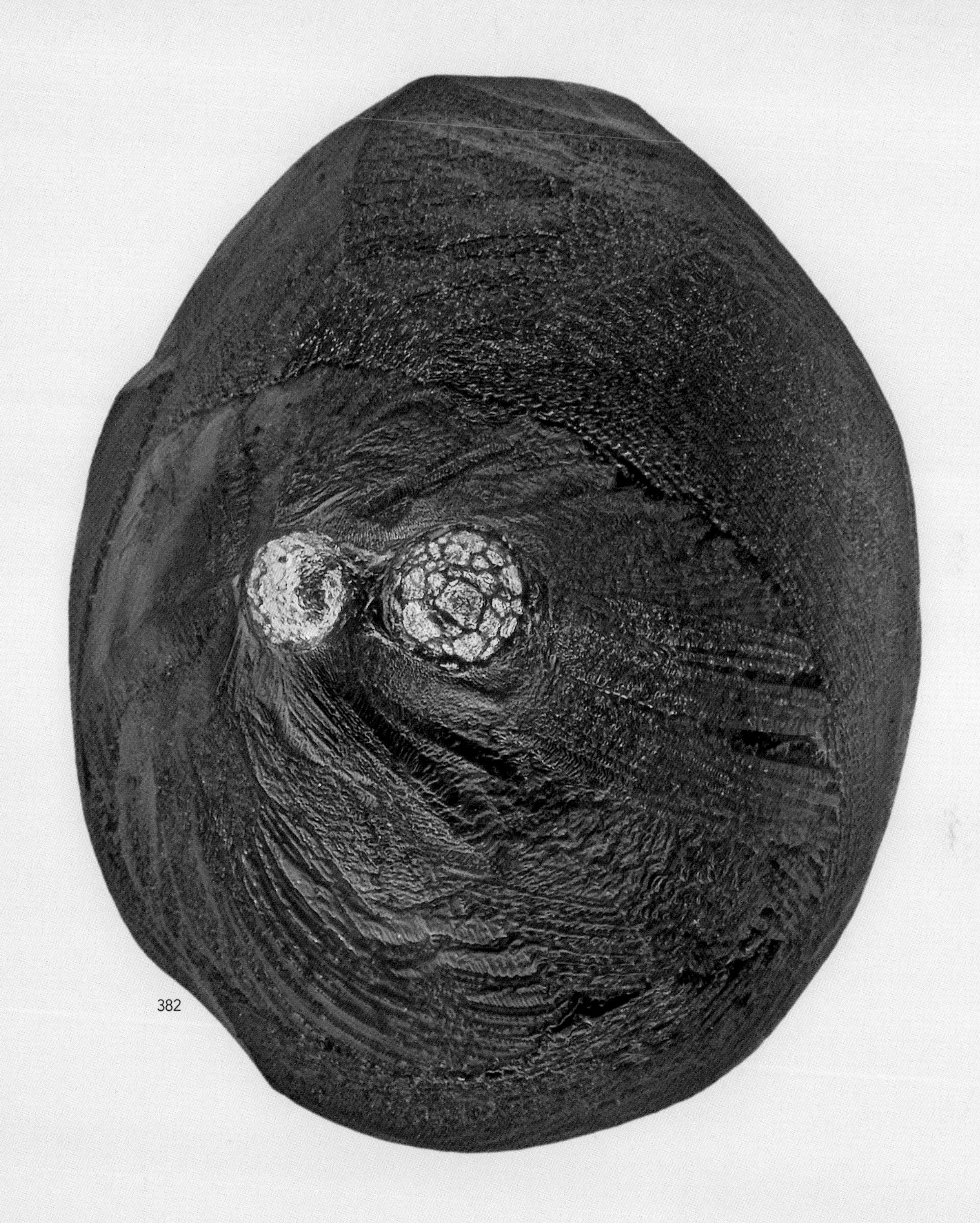
382

365

SEM SECTION DETAILS

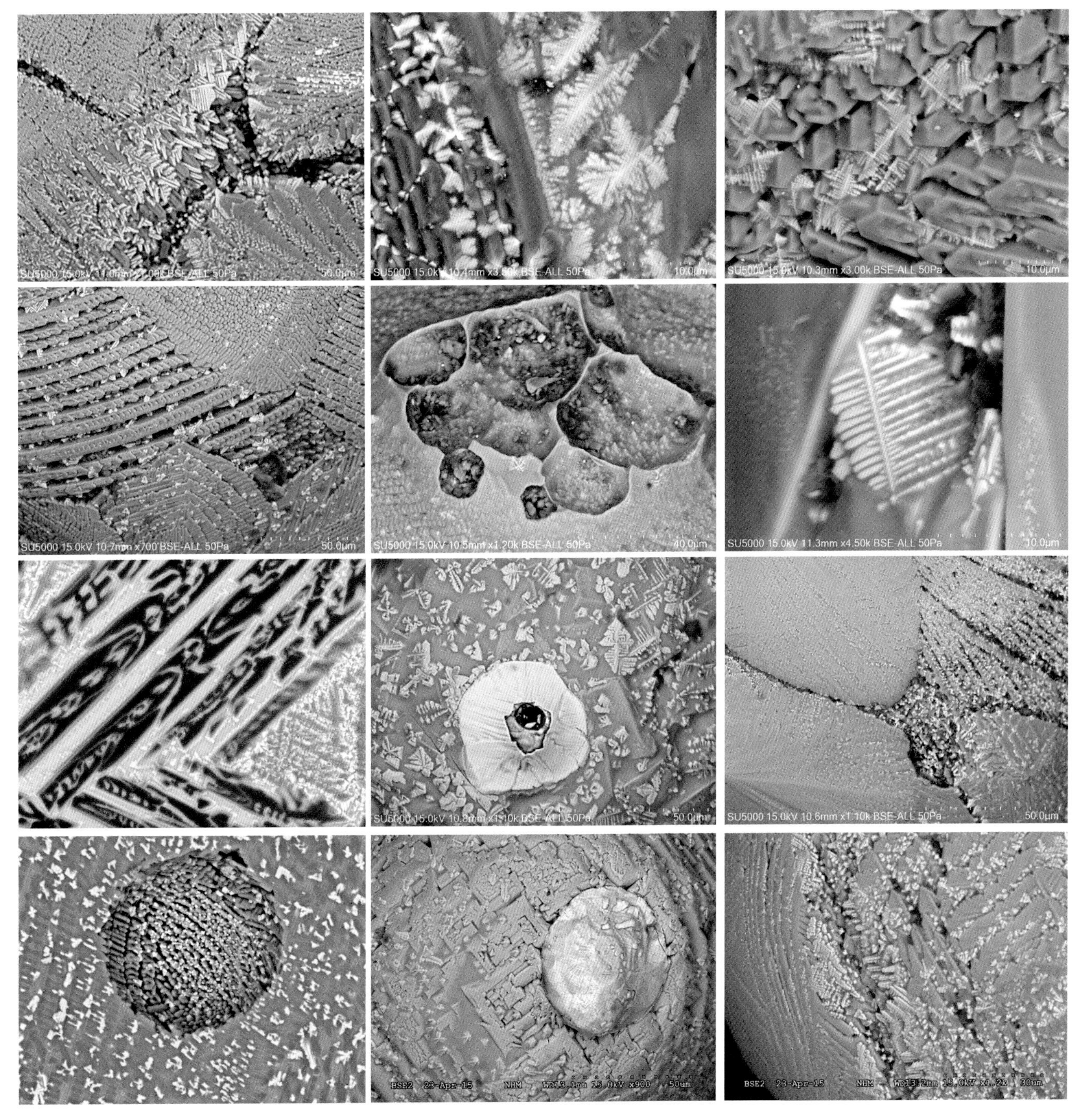

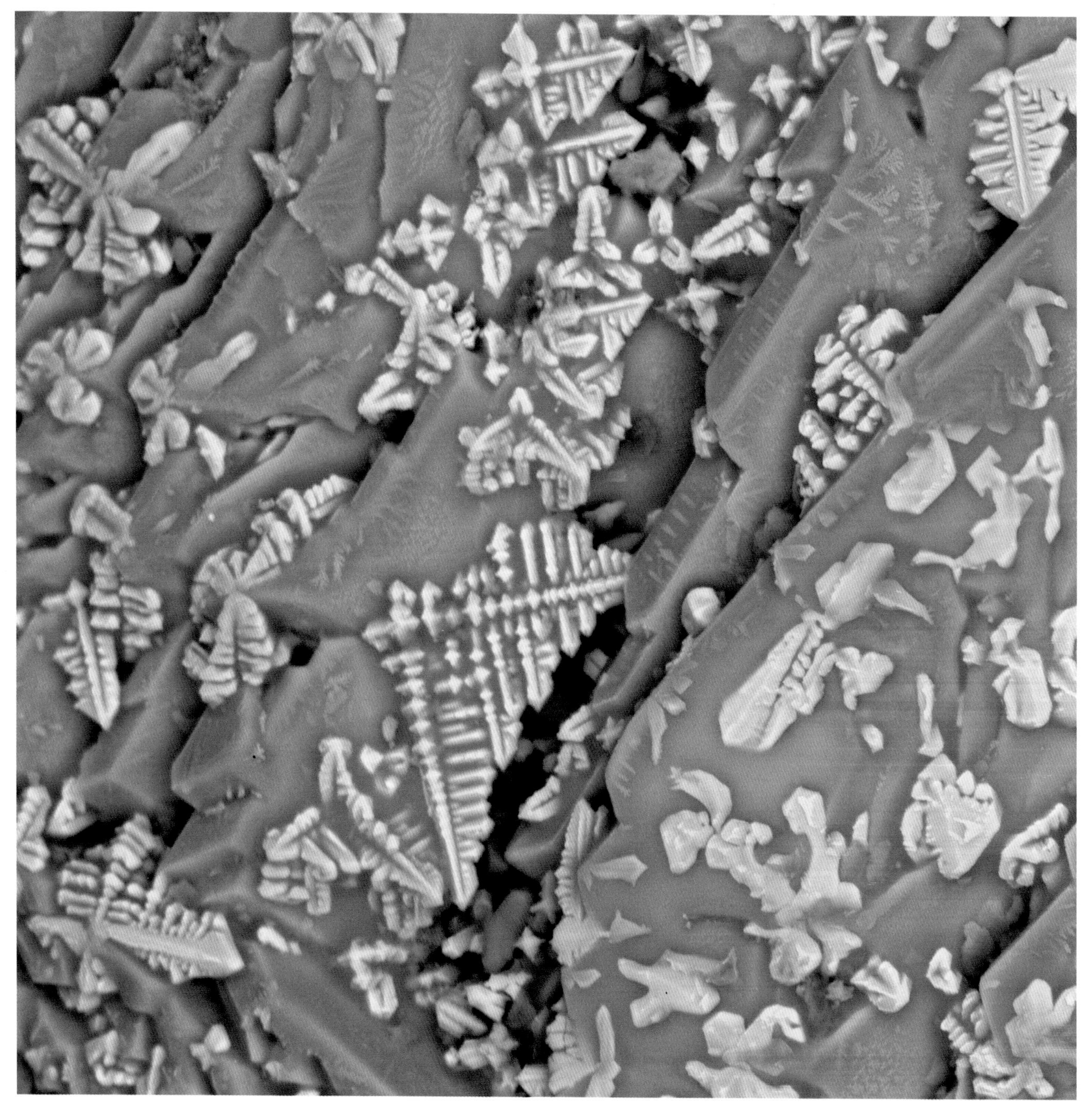

458

505

SEM SECTION DETAILS

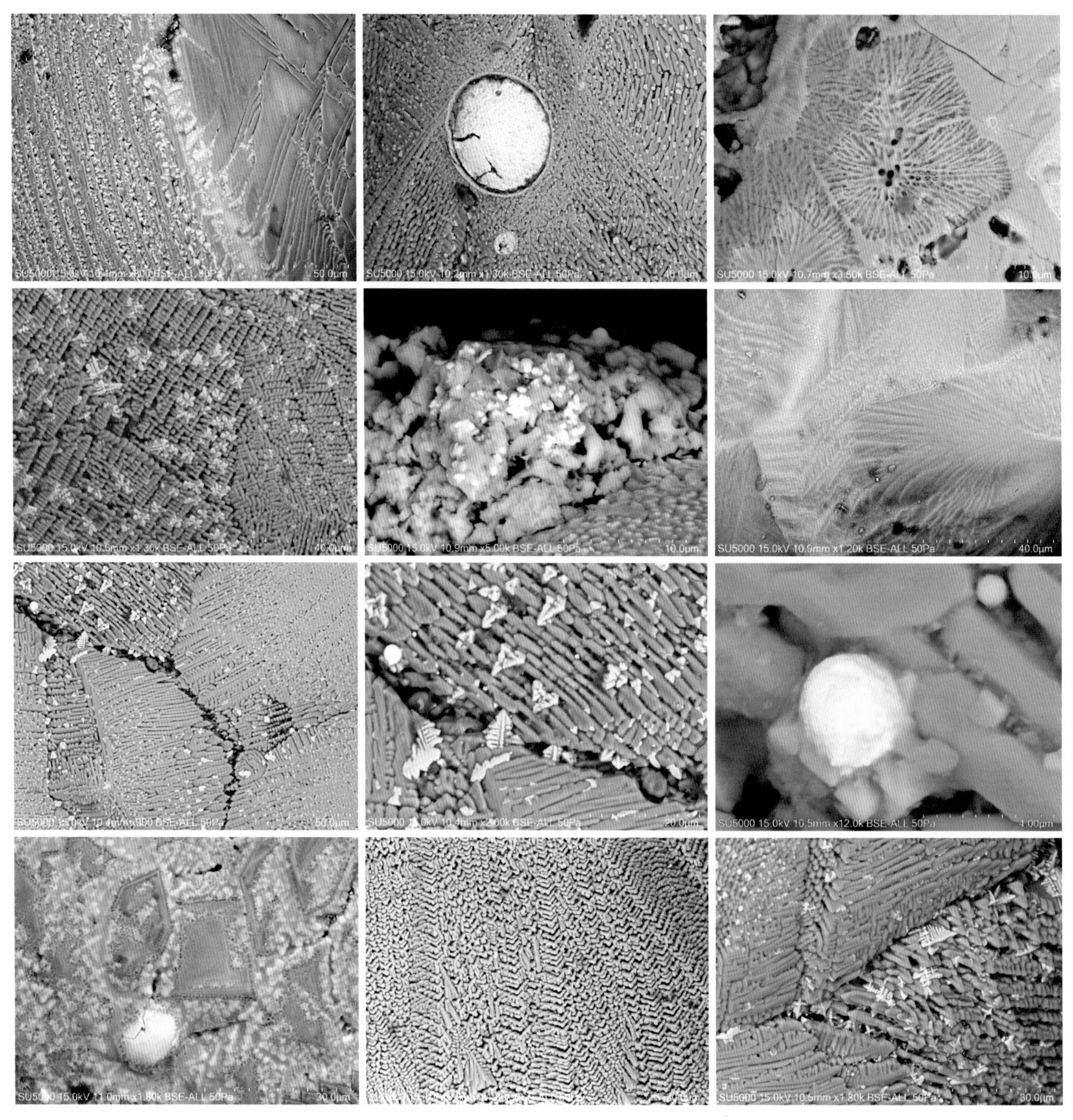

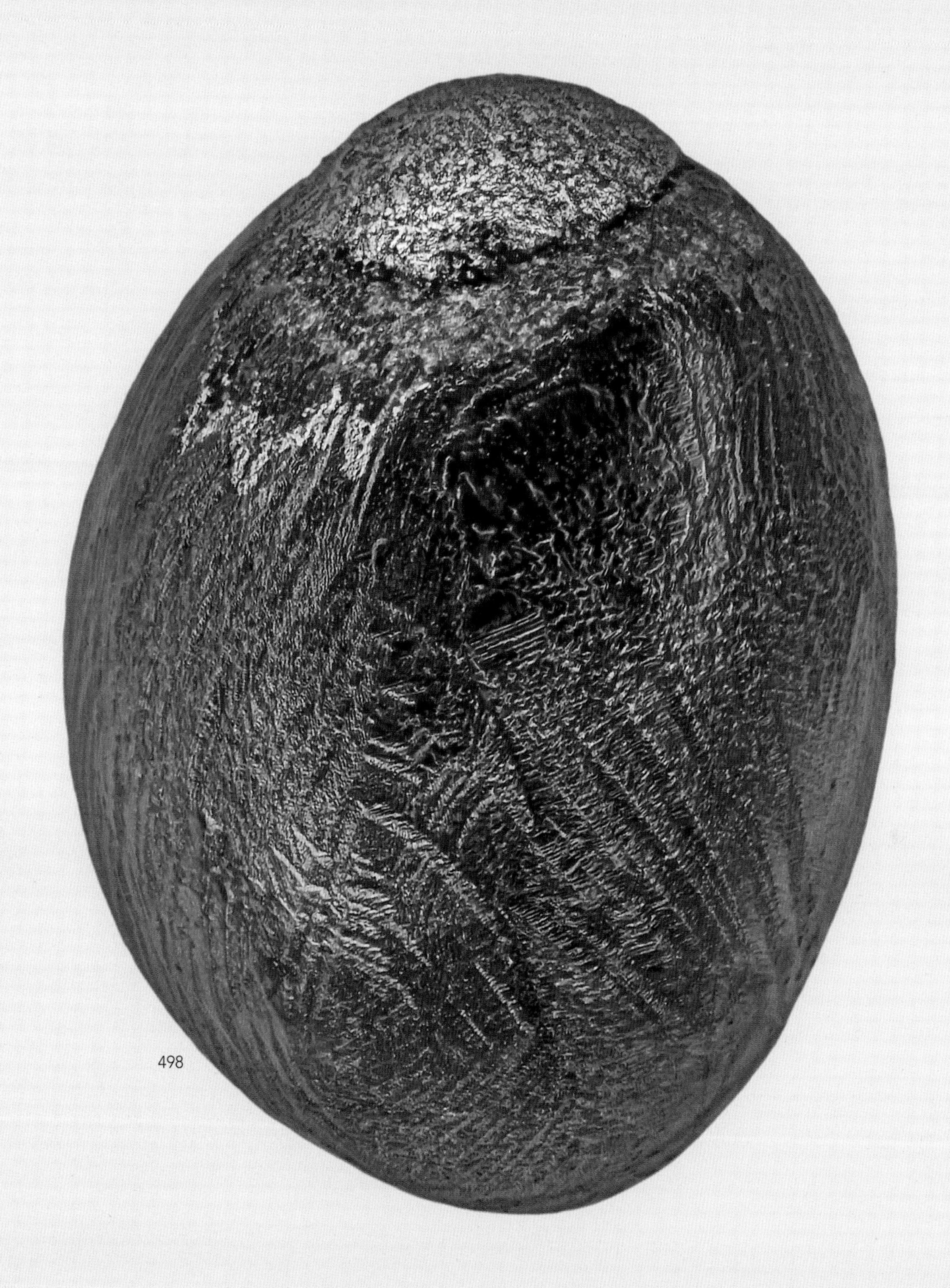

498

Ablation Spherules

It has been calculated that an average meteoroid loses around 85% of its mass during atmospheric flight. Some of the eroded mass ends up as ablation spherules.

The term ablation spherule (a melted spheroid micro object rubbed off a meteoroid by atmospheric friction) is often used to describe just about any micro spherule found on the ground, possibly to avoid dubious use of the word micrometeorite. But the ablation spherules are not true micrometeorites despite their extraterrestrial origin, since they were not small in space. They are more closely related to the meteoritic fusion crust.

The ablation spherules (~0.1–0.2 mm) on these pages are from the Chelyabinsk event that occurredon February 15, 2013, in Russia. They appeared as black dust on the fresh snow, and it is estimated that 12,000–13,000 metric tons (greater than 99.99%) of the large meteoroid suffered ablation in the atmosphere. The dust plume then unexpectedly streamed back upwards into the stratosphere by the jet streams, and within seven days the cloud of ablated particles covered the entire Northern Hemisphere before the spherules eventually fell to the ground.

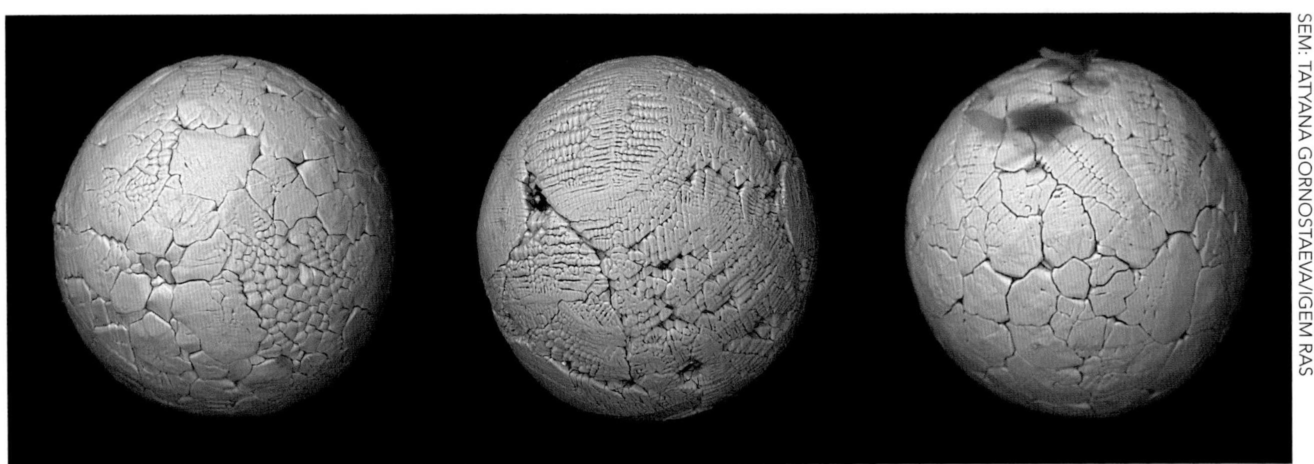

SEM: TATYANA GORNOSTAEVA/IGEM RAS

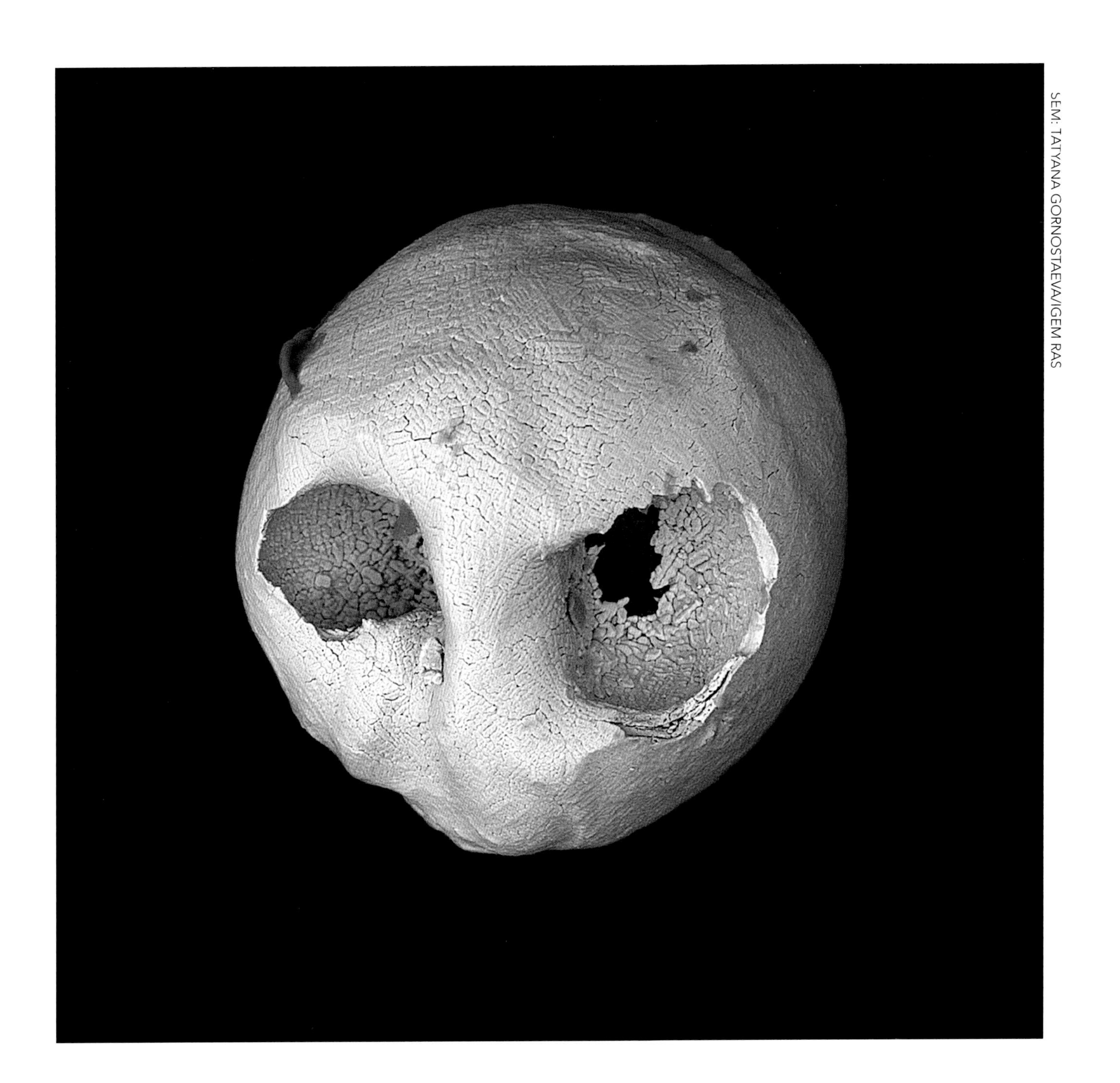

SEM: TATYANA GORNOSTAEVA/IGEM RAS

The Enigmatic Chondrules

At 12:30 a.m. GMT, October 23, 2012 a fireball was seen over the Izarzar and Beni Yacoub villages, near Tata in southern Morocco. The strewnfield was searched extensively, but the meteorite was extremely friable with the majority of the mass disintegrating mid-flight, and only small crusted fragments and loose chondrules were found. Twenty-two of these (~0.8–3.0 mm), collected within days of the fall, are shown on page 78.

Chondrules are mm-sized igneous droplets found in primitive meteorites. They formed in flash heating events in the Solar Nebula about 4.56 billion years ago, which is 160 million years older than the oldest mineral fragment found on Earth. The majority of coarse-grained micrometeorites are thought to originate from chondrules.

The chondrules on these pages are from the following meteorites: BJURBÖLE (Finland 1899, classified as L/LL4, below, top row), VALLE (Norway 2013, H-chondrite, below bottom row), NWA 5929 (Northwest Africa 2009, LL5, pages 80–81), and IZARZAR (Morocco 2012, H5, page 78). Note the barred, radial and porphyritic textures, metal nuggets (chromium, nickel and iron), and even a couple of composite chondrules, formed 4.56 trillion years ago.

CHONDRULES

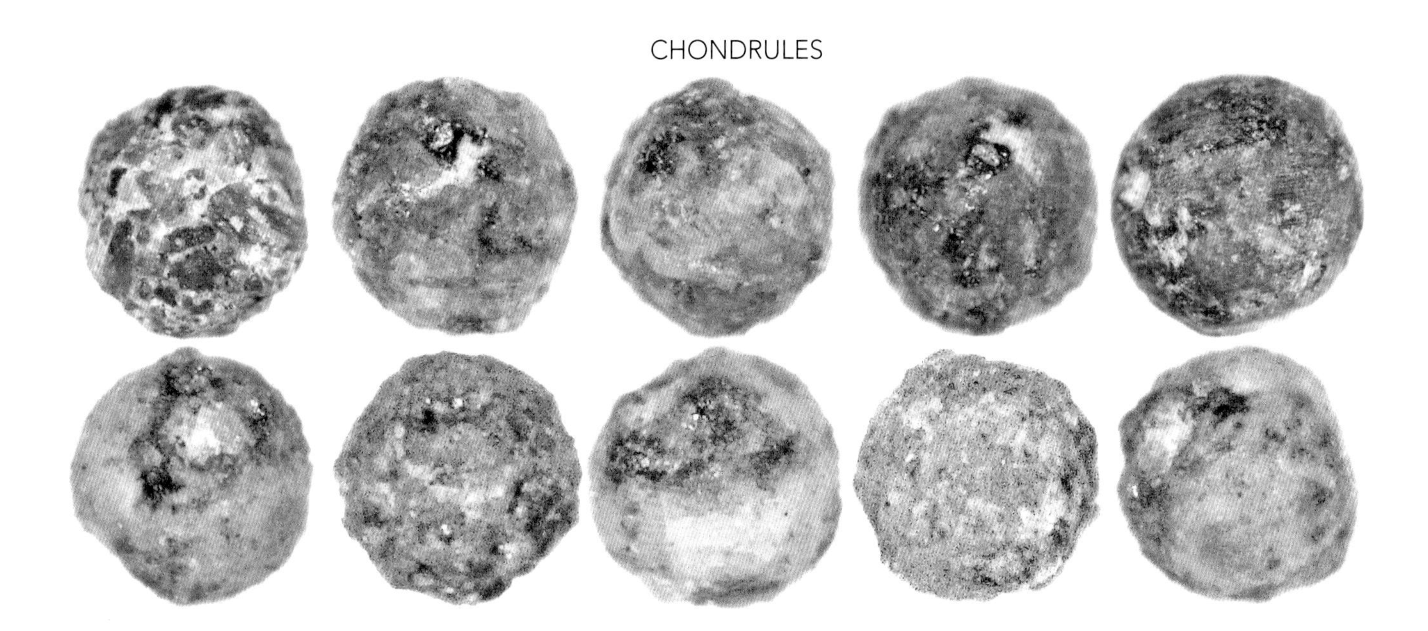

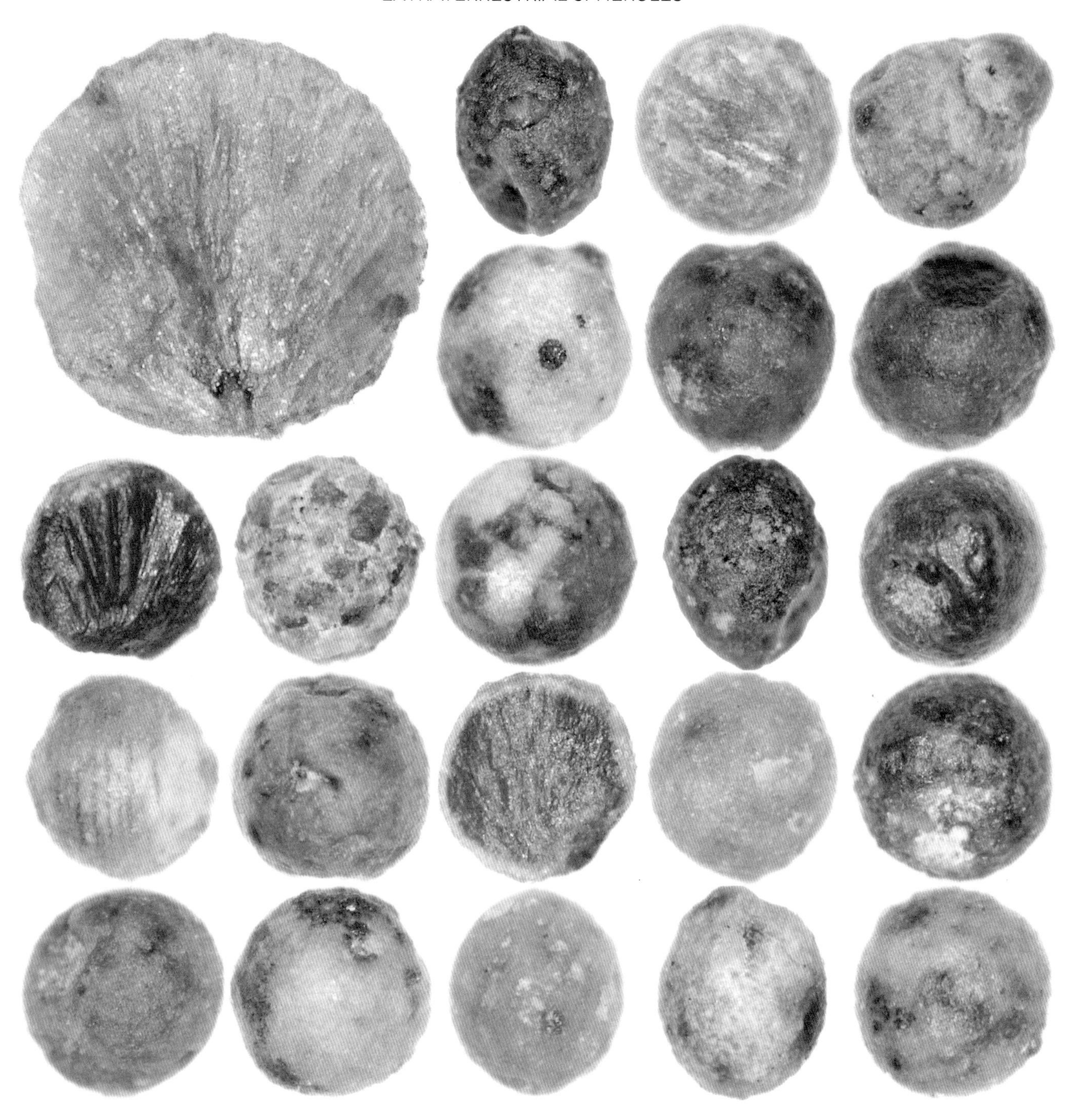

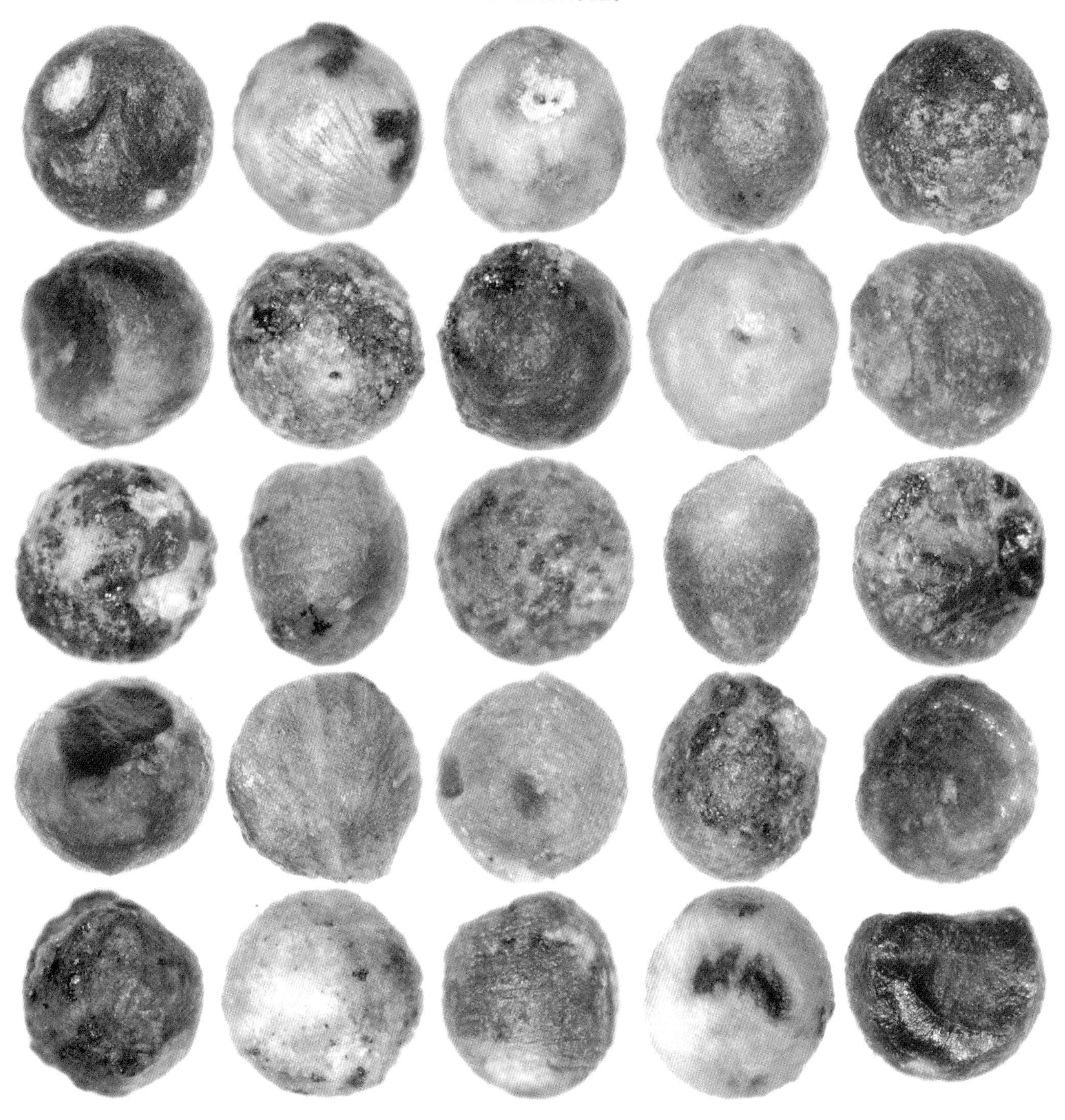

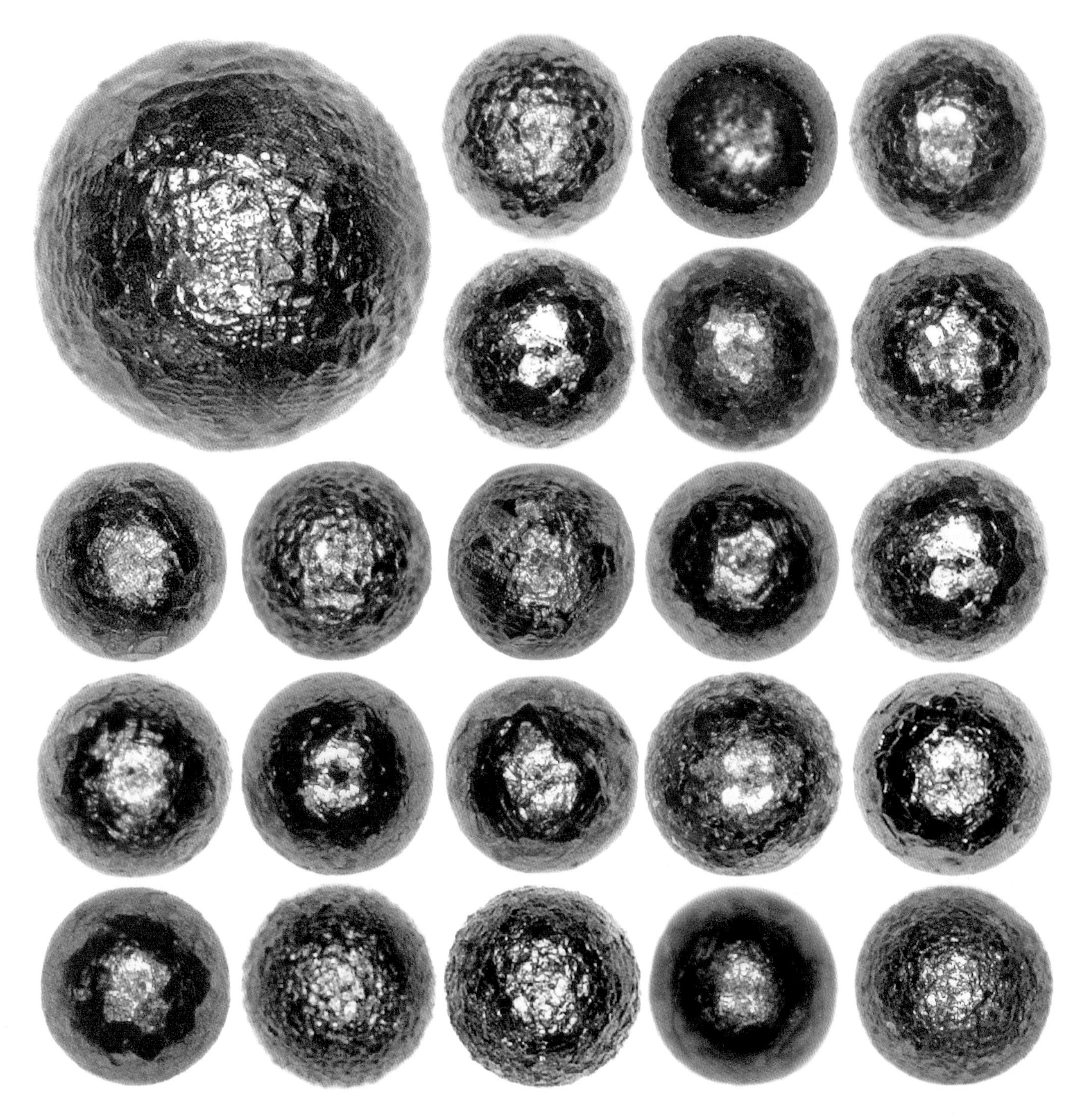

III ANTHROPOGENIC SPHERULES

I-Type Magnetic Spherules

Magnetic I-type cosmic spherules are iron oxides, mainly magnetite (see pages 124–125) and wüstite. They are resistant to weathering, but rare–amounting to only around 2% of the total number of melted micrometeorites. In the deep sea collections, however, they occur abundantly, as a result of the lower weathering resistance of the stony micrometeorites.

In the search for magnetic micrometeorites in urban areas we find large numbers of I-type spherules. It is unproblematic to distinguish them from naturally occurring minerals, but they are not extraterrestrial; their origin is anthropogenic. Many mechanical and industrial processes, and all sorts of power tools—oxy fuel-cutting torches, grinding wheels, angle cutters, etc.—produce I-type spherules, and they are distributed everywhere by the wind, rain, and human activities.

In the case of the I-type spherules a chemical EDS analysis cannot reveal the provenance unless additional clues like nickel beads or platinum group nuggets can be found. Consequently the I-type spherules found in populated areas, like the ones on these pages, can be considered anthropogenic, unless verified extraterrestrial.

I-TYPE MAGNETIC SPHERULES

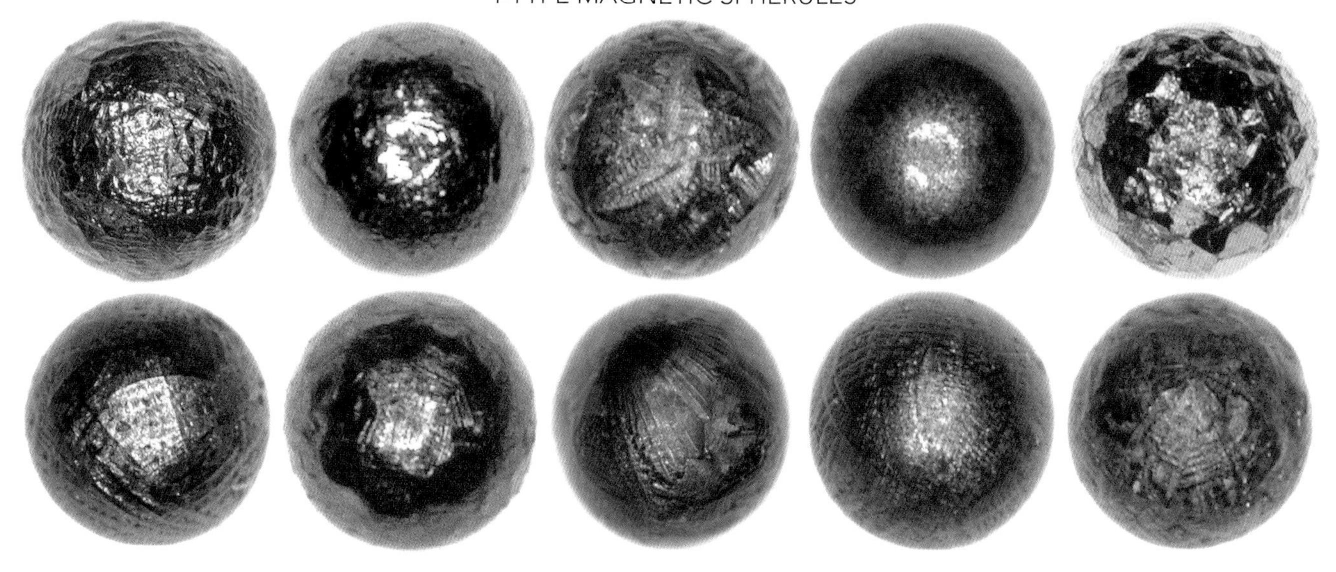

I-TYPE MAGNETIC SPHERULES

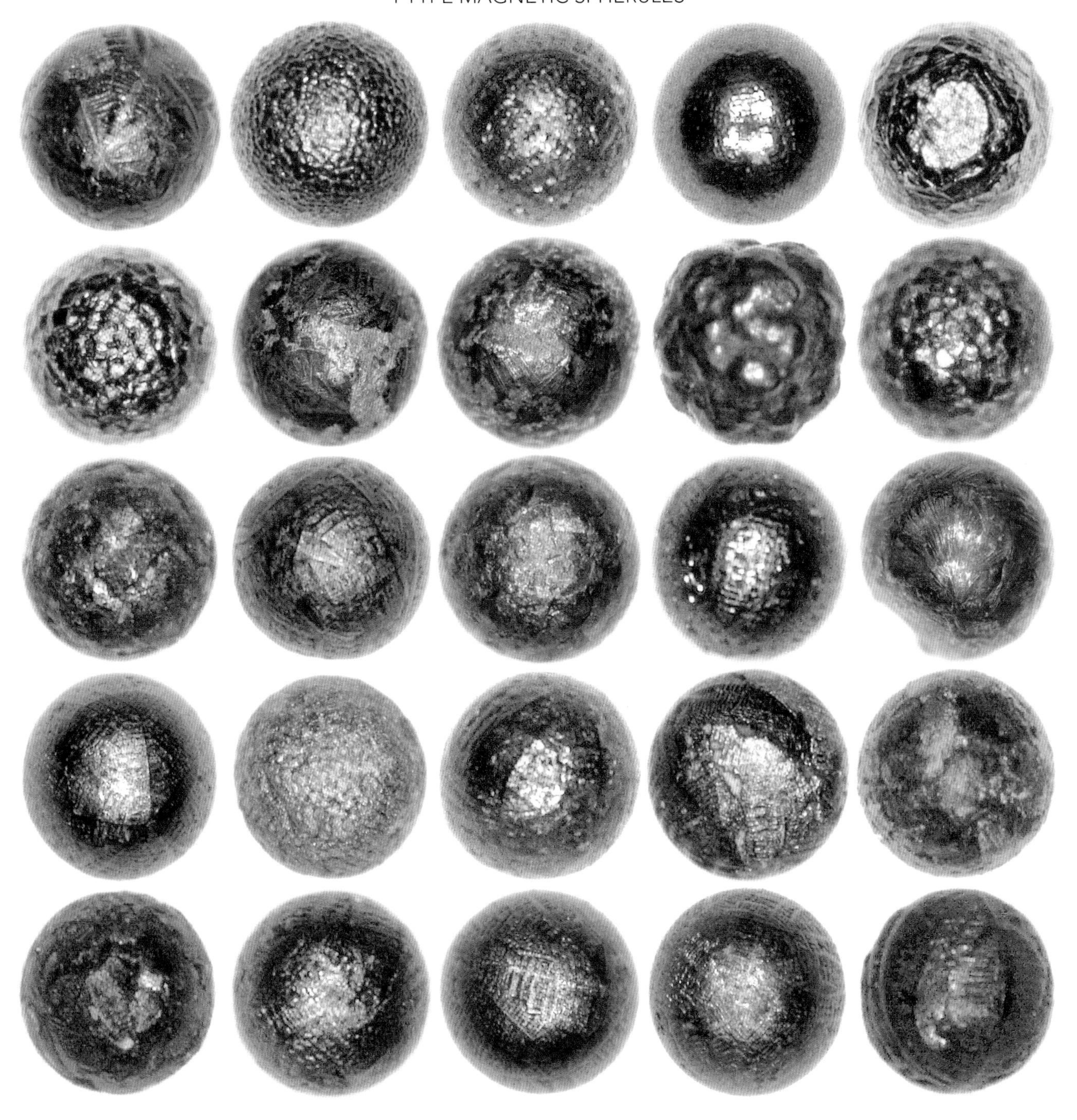

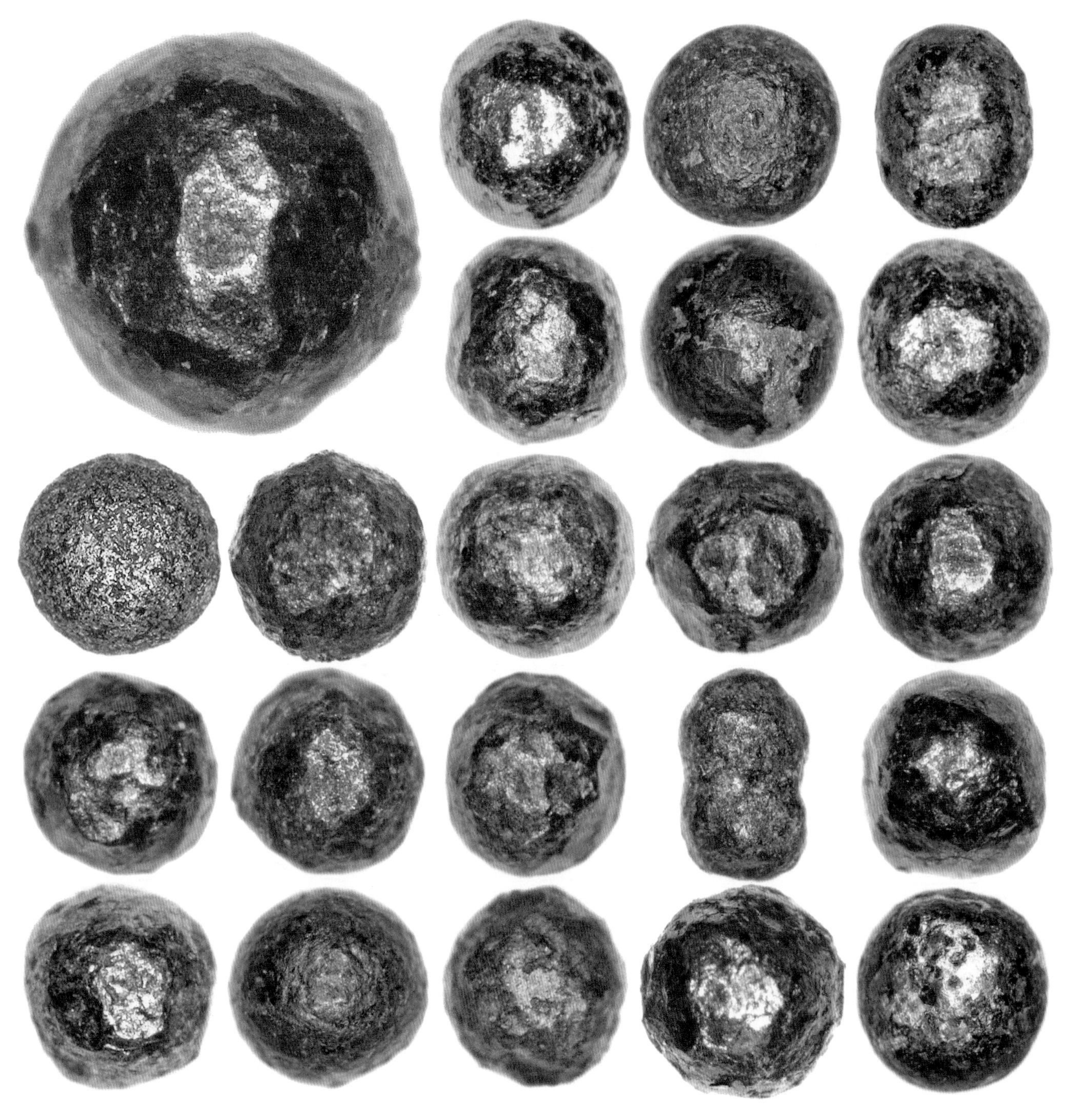

Massive Iron Spherules

The regular I-type (iron) spherules, both extraterrestrial and anthropogenic, may contain a central void due to rapid solidification inward from the surface. This is not the case with the massive I-type spherules, which also have a different morphology than the regular I-type, albeit the chemical spectrum is the same: iron oxides. They are often slightly elongated, sometimes with rudimentary polygonal faces. Unlike the regular I-type spherules, the massive I-type rusts quickly, appearing like small rusty cannon balls.

The exact origin of these 0.5–1.0 mm spherules is uncertain, but definitely terrestrial. They show no sign of ablation, etc. The sheer number to be found along roads, especially in curves and steep hills, may point toward vehicles, perhaps the brake systems of heavy trucks, but this is merely speculation. The point is that the massive I-type spherules are not ET, but it is necessary to know them well enough to recognize and disregard them in the search for micrometeorites in road dust.

Nuggets, Beads, and Cores

Platinum group nuggets (PGNs), nickel/chromium containing beads and other types of metal inclusions are characteristic features in approximately 5% of all cosmic spherules. In size they vary from submicron-sized PGNs to large nuggets, or cores. Some micrometeorites have a hole where a metal bead has escaped (see page 67) so it should also be possible to find loose nickel/iron cores.

The differentiation where the heavier elements sink inward and form a core happens rapidly when the particle is in a liquid state, and the surface tension gives the spheroid form. This also happens with anthropogenic spherules like the ones on these pages. Note the lack of aerodynamic ablation and absence of magnetite "Christmas trees" compared with the micrometeorites in the first part of the book.

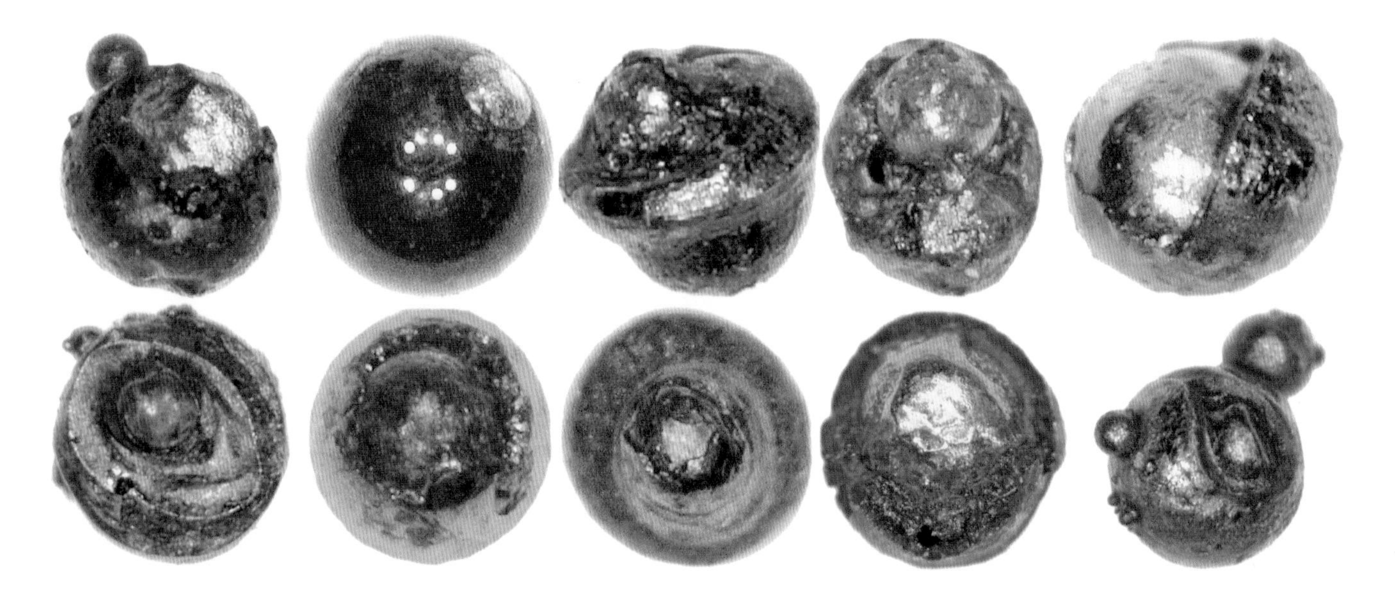

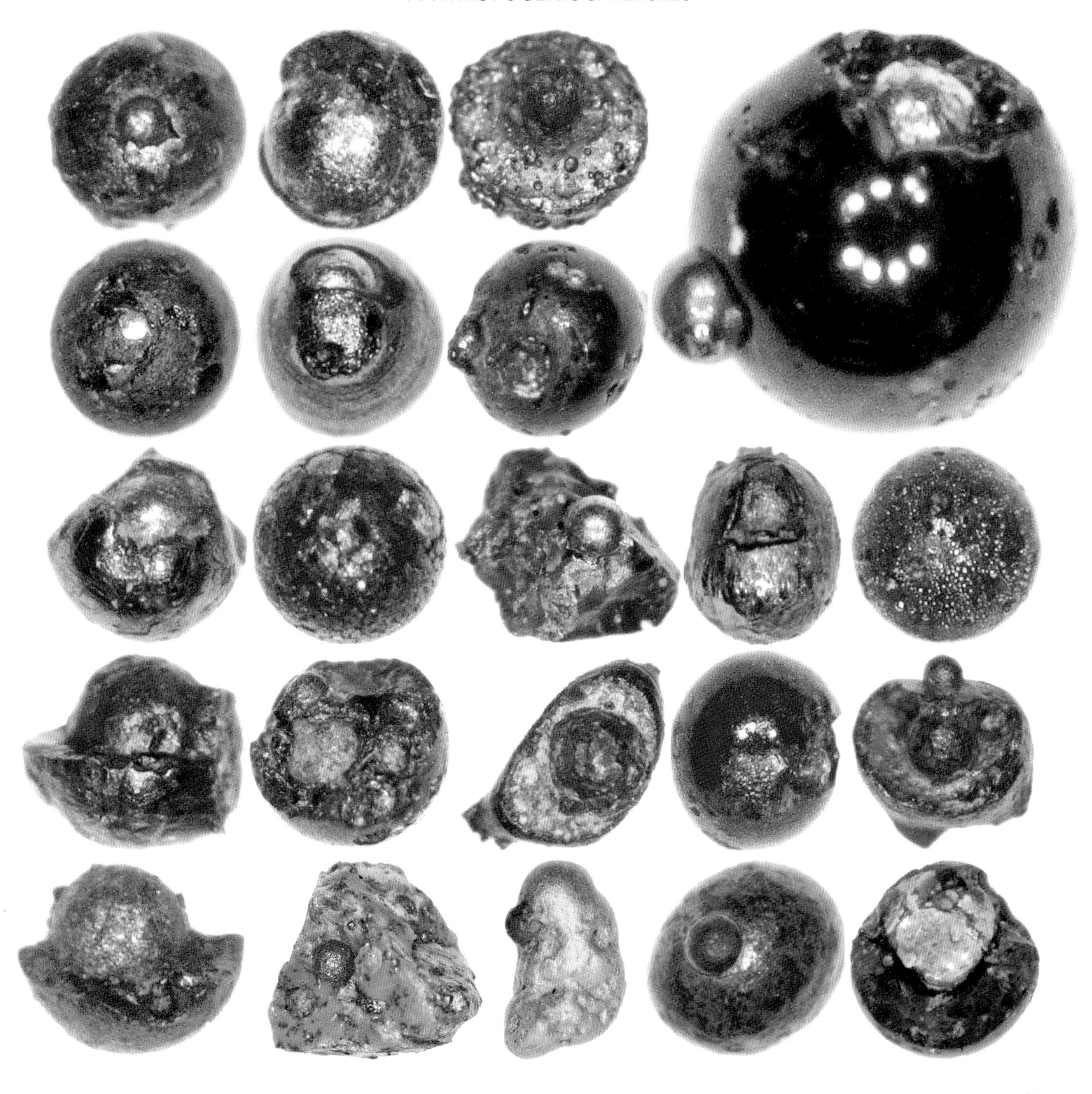

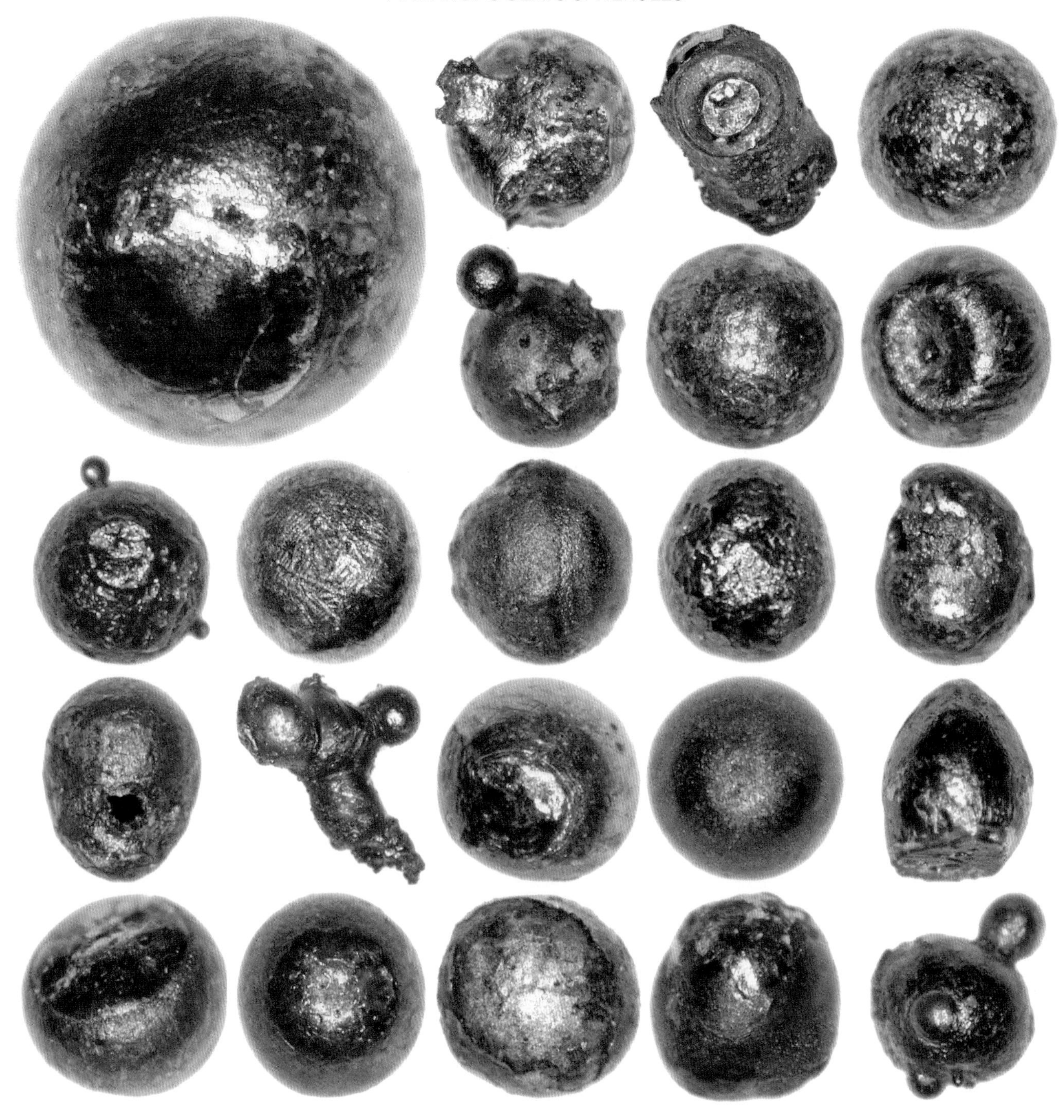

From the Welding Shop

The main challenge in the search for micrometeorites in populated areas is to distinguish the extraterrestrial particles from the terrestrial. Here we can take a lesson from the 2,500-years-old Chinese book *The Art of War*, which advises "know your enemy." Inspired by this ancient wisdom, I went to a welding shop and asked to sweep the floor.

Subsequent examination of the dust samples under a binocular microscope revealed a panorama of magnetic spherules with a variety of morphological features dominated by composite particles, twins, splash and crash formations—which rules out a micrometeoritic origin. Although we cannot know exactly which spherule originates from which power tool, we can recognise the anthropogenic "signatures," and narrow the possibly extraterrestrial candidates.

Sparks

In a not-too-distant past when smoking cigarettes was common, it was assumed that looking for micrometeorites in populated areas would be impossible because each lighter spark from lighting a cigarette created spherules indistinguishable from micrometeorites.

The spherules on this page were created by igniting a lighter so the sparks hit a carbon plate used for scanning electron microscope examination. Numerous spherules were found, but they do not have chondritic chemistry, and note the scale bar and size of the spherules: a maximum of ~5-6 micron. This is 1/50 of an average micrometeorite, and 1/10 of the practical lower limit of spherules in MM collections. By excluding spherules less than 50 microns, many contaminants are avoided.

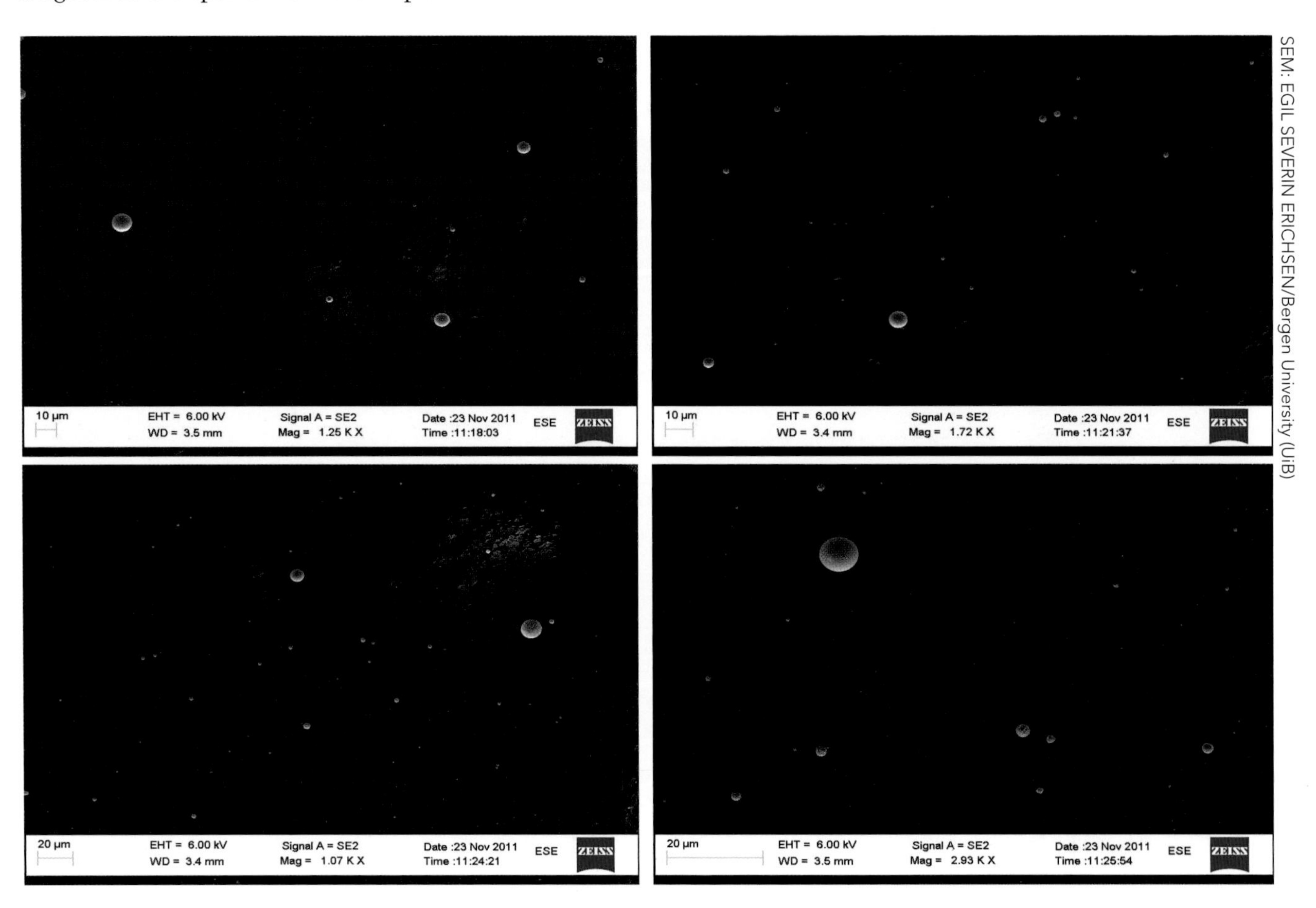

SEM: EGIL SEVERIN ERICHSEN/Bergen University (UiB)

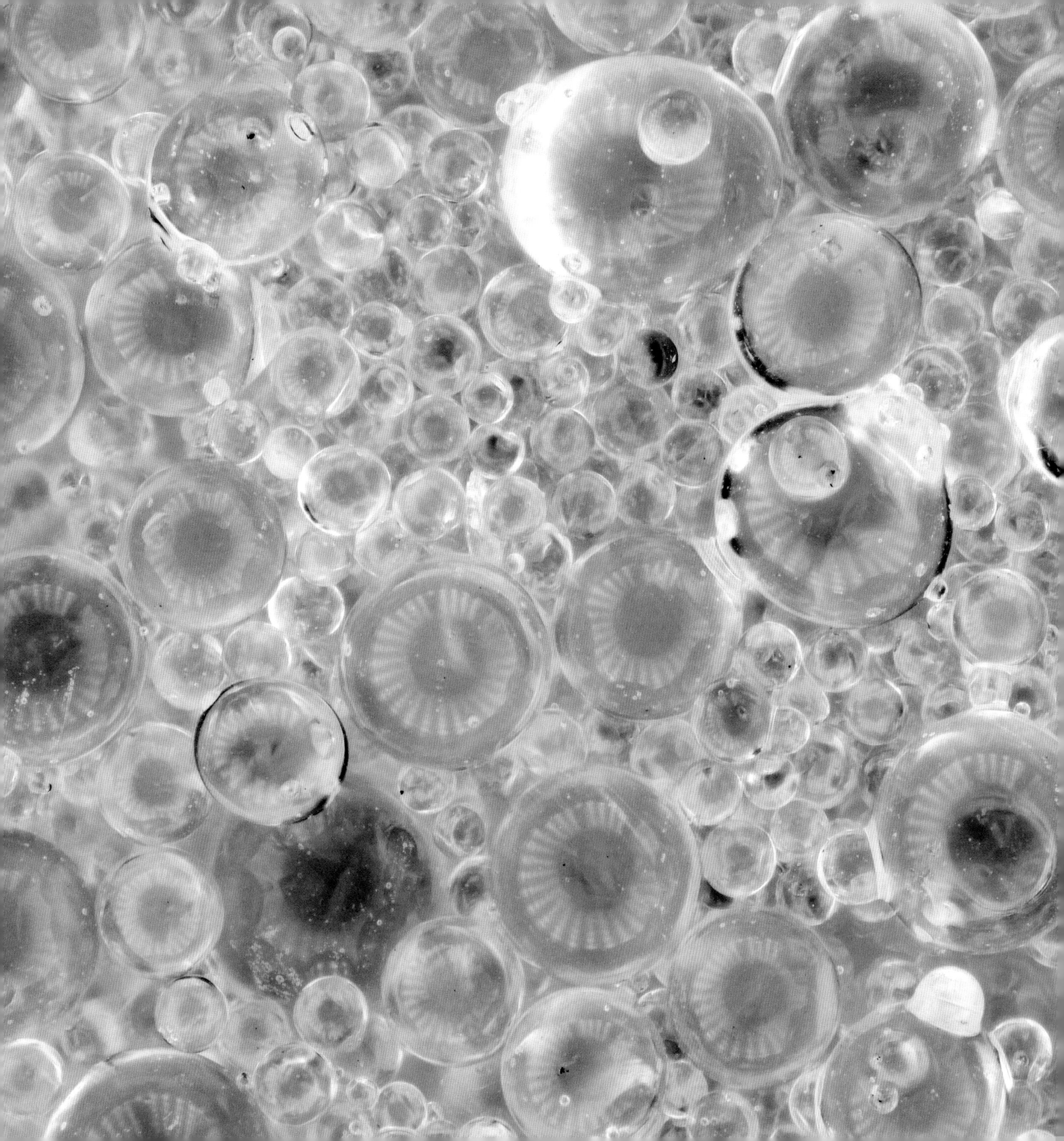

Nonmagnetic Glass Spherules

Up to 98% of the total number of cosmic spherules are stony, including the completely melted glass spherules. These V-type (vitreous) micrometeorites are usually spherical and transparent, but can be highly vesiculated. They are colourless to brown or green and normally nonmagnetic unless they contain nickel-iron beads (see pages 46 and 51).

In populated areas all over the world the most common type of all spherules are the omnipresent nonmagnetic glass spherules from road surface marking (see page 93). These retroreflective glass beads are produced industrially of flawless colorless silica glass in defined fractions and sometimes they constitute the main part of urban road dust. There is also a wide range of anthropogenic glass spherules produced by power tools, industry and human activities. These may have colors and vesicles reminiscent of the cosmic glass spherules, but their terrestrial origin is often revealed by composite forms, multiple tails, etc.

So far it has not been possible to separate the cosmic glass spherules from the anthropogenic ones by morphology alone—a chemical EDS analysis is neccessary to verify the chondritic spectrum and ET origin. Consequently, as with the I-type spherules (see page 83) all glass spherules in populated areas can be considered anthropogenic unless verified as extraterrestrial.

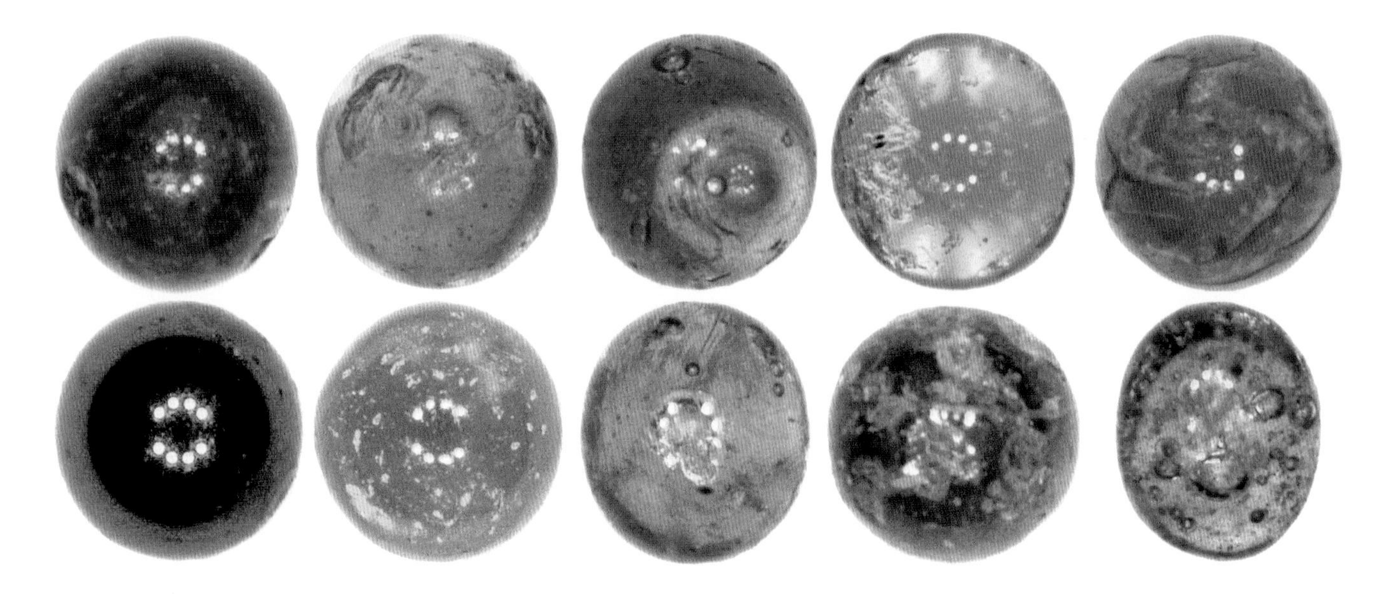

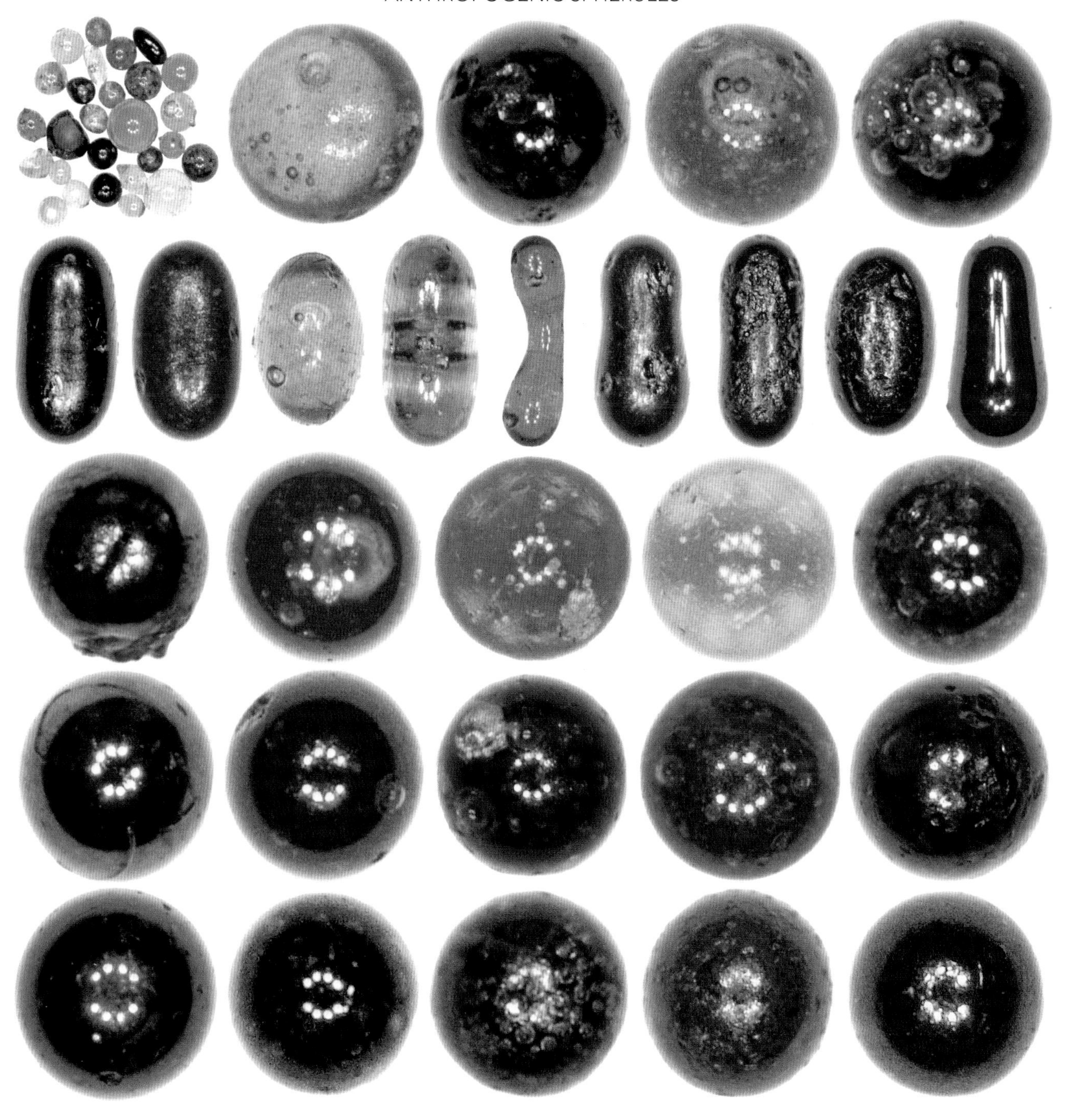

Spherules from a Steam Locomotive

There is a widespread misconception that searching for micrometeorites in the vicinity of railroad lines is impossible, due to spherules produced by the trains. In the early days of the MM research the contamination from steam locomotives may have discouraged more than one good scientist, but today these veteran trains are rare.

The photos on pages 96–98 are of micro objects found in the boiler and the smoke box of an old steam locomotive, Tertitten, Sørumsand, Akershus, Norway. The spherules are produced by small mineral impurities in the fuel, in this case high-quality Polish coal. They display an impressive variation in morphology.

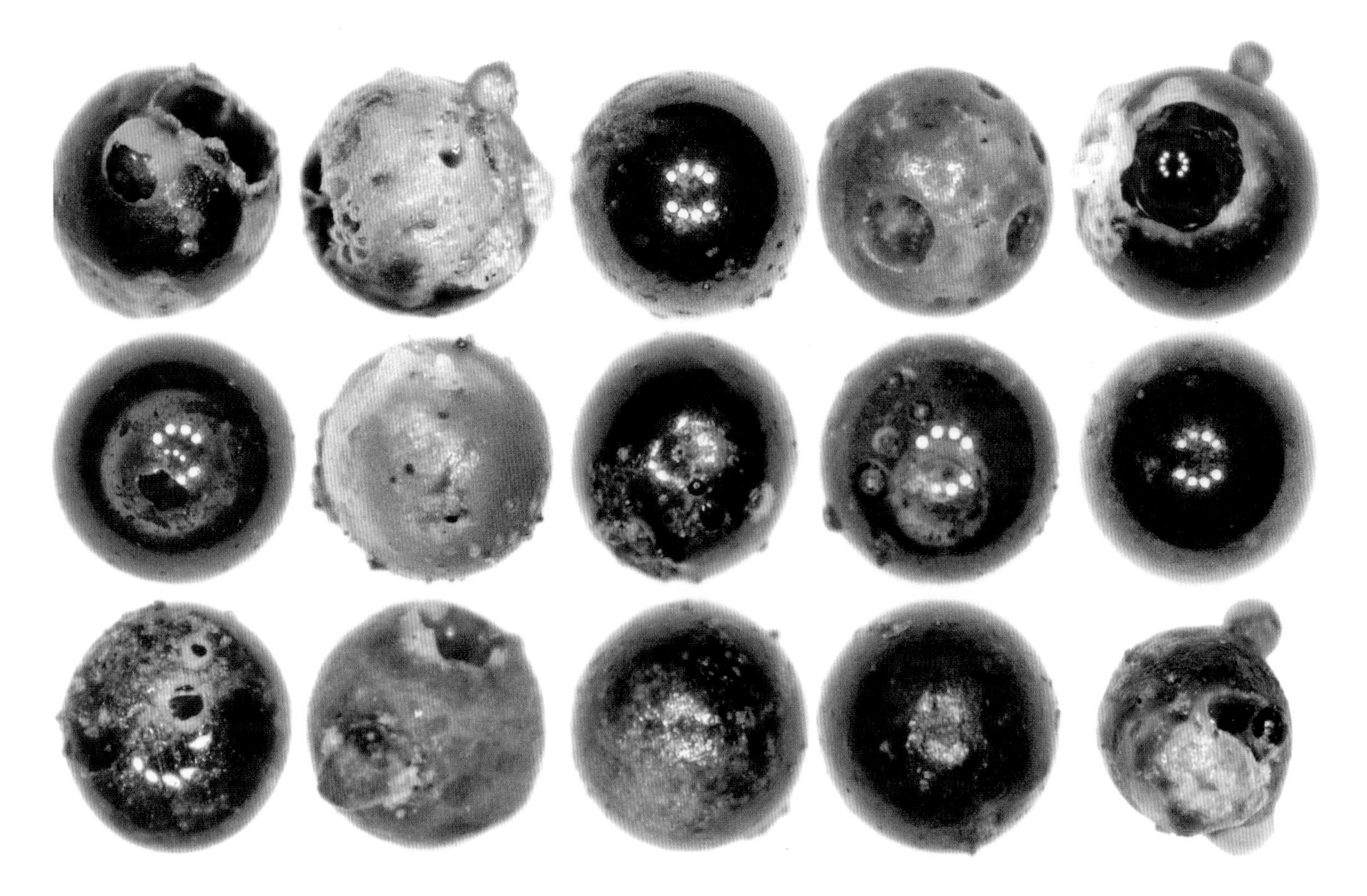

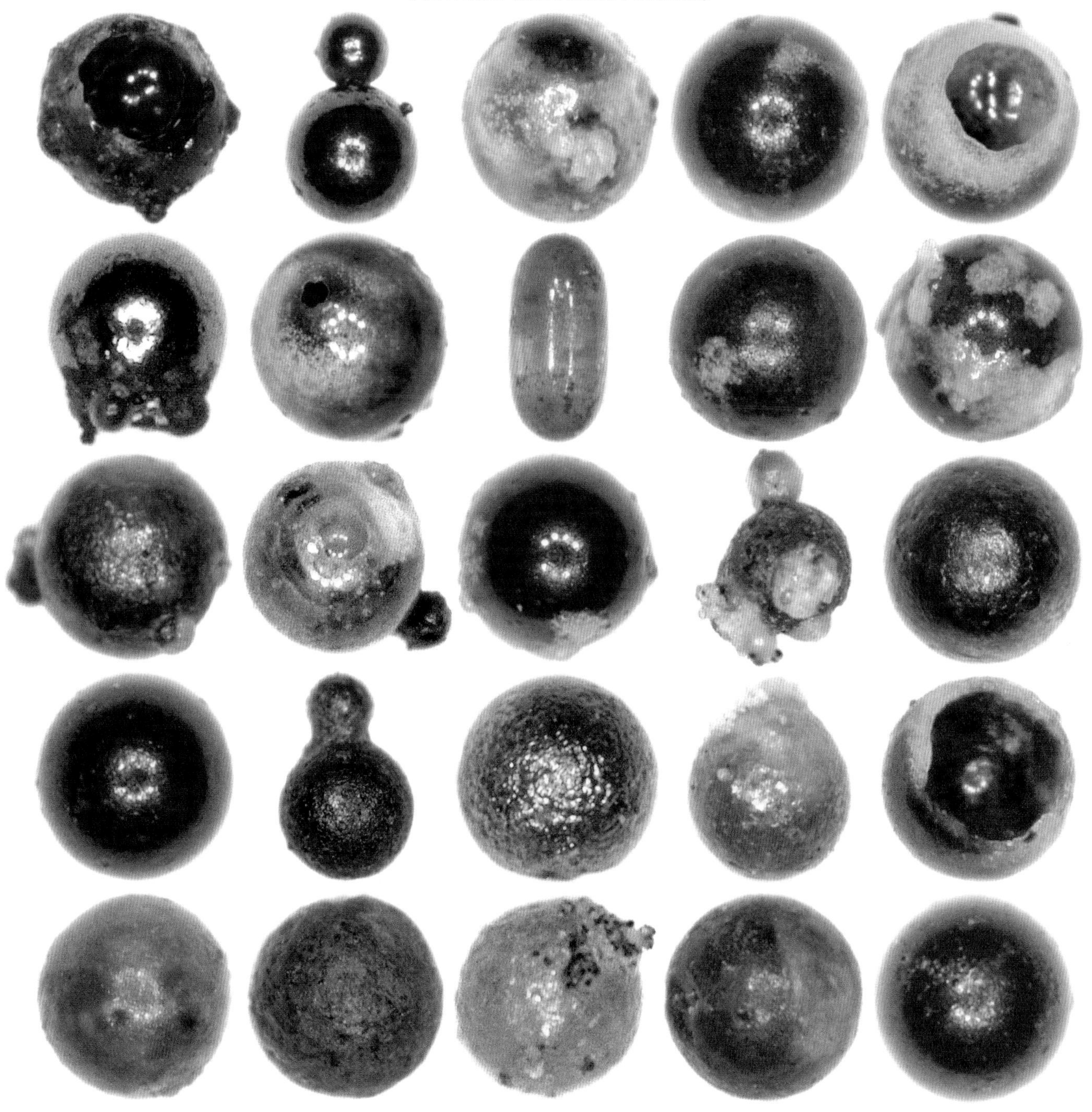

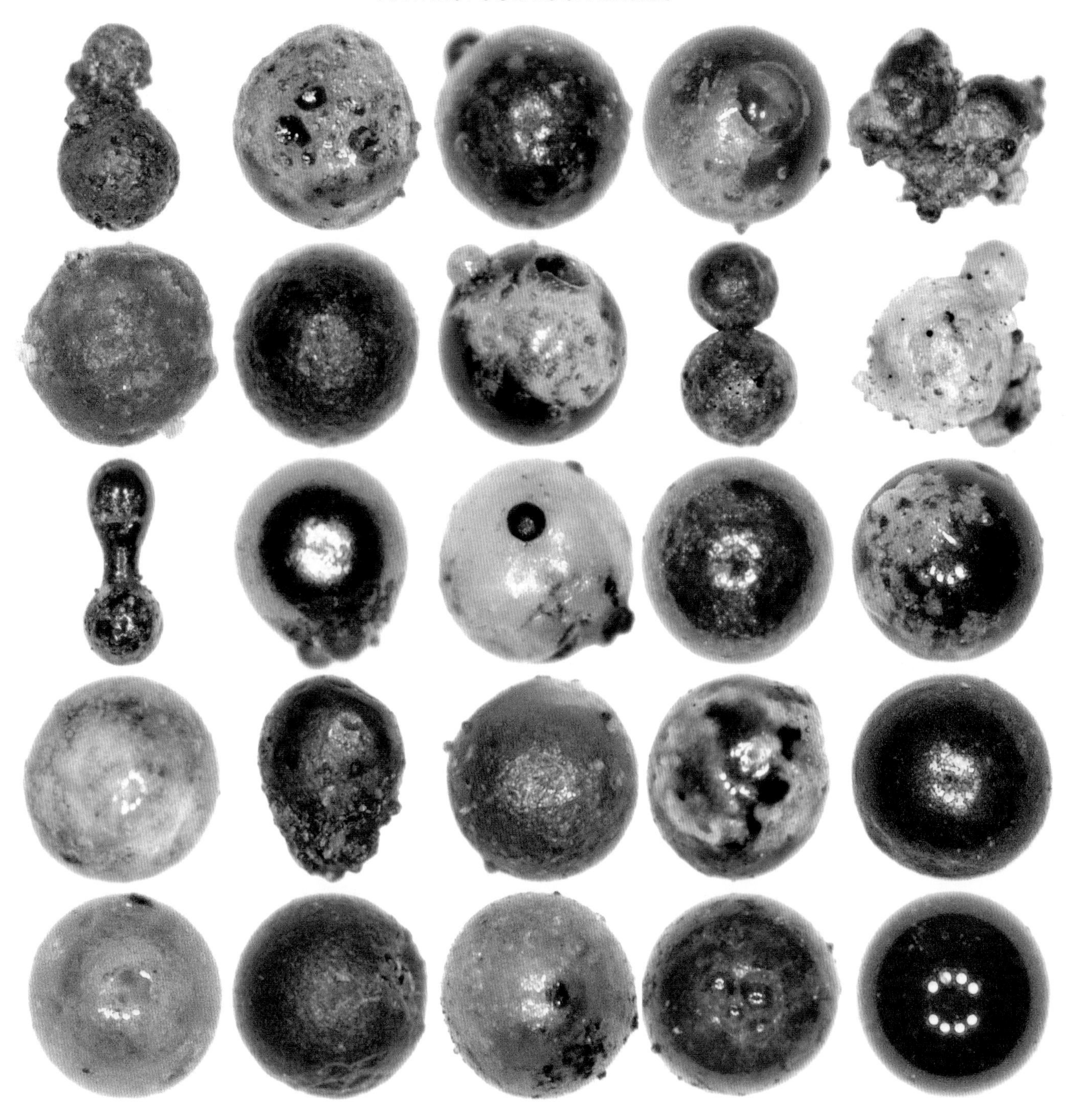

Mineral Wool

Mineral wool like the Rockwool particles on this page is produced from basalt and chalk. The rocks are heated to 1,600 °C (2,912 °F) into a melt, which is blown into a large spinning chamber, which pulls the melt into fibers. At the end of a fiber a droplet may form, and spherules are common, often with one or more tails. The spherules are nonmagnetic and have terrestrial basaltic chemistry.

On a field trip to Moss town (Norway), I had the opportunity to examine dust samples from the rooftop of Kirkeparken Highschool, where Harald Kolderup and Anders Rekaa have built a collecting device for micrometeorites. Three km away is a Rockwool factory, and despite their particle emission filter, these 0.2–0.5 mm Rockwool spherules were found on the roof, possibly carried by the wind.

Mineral wool is used worldwide as insulation, and droplets and spherules like these are found in the most unexpected places. A simple chemical analysis will reveal if it is chondritic—a micrometeorite.

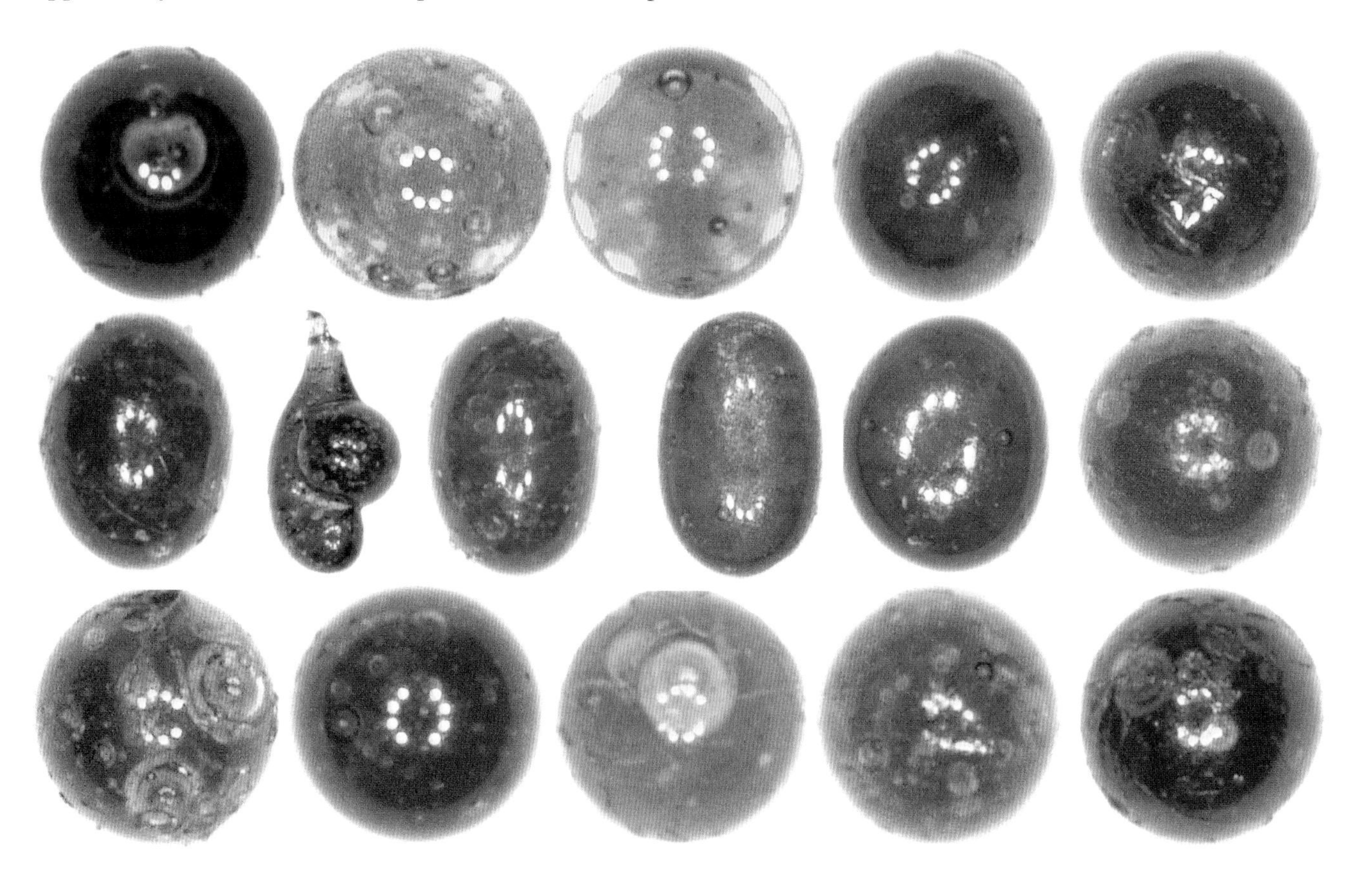

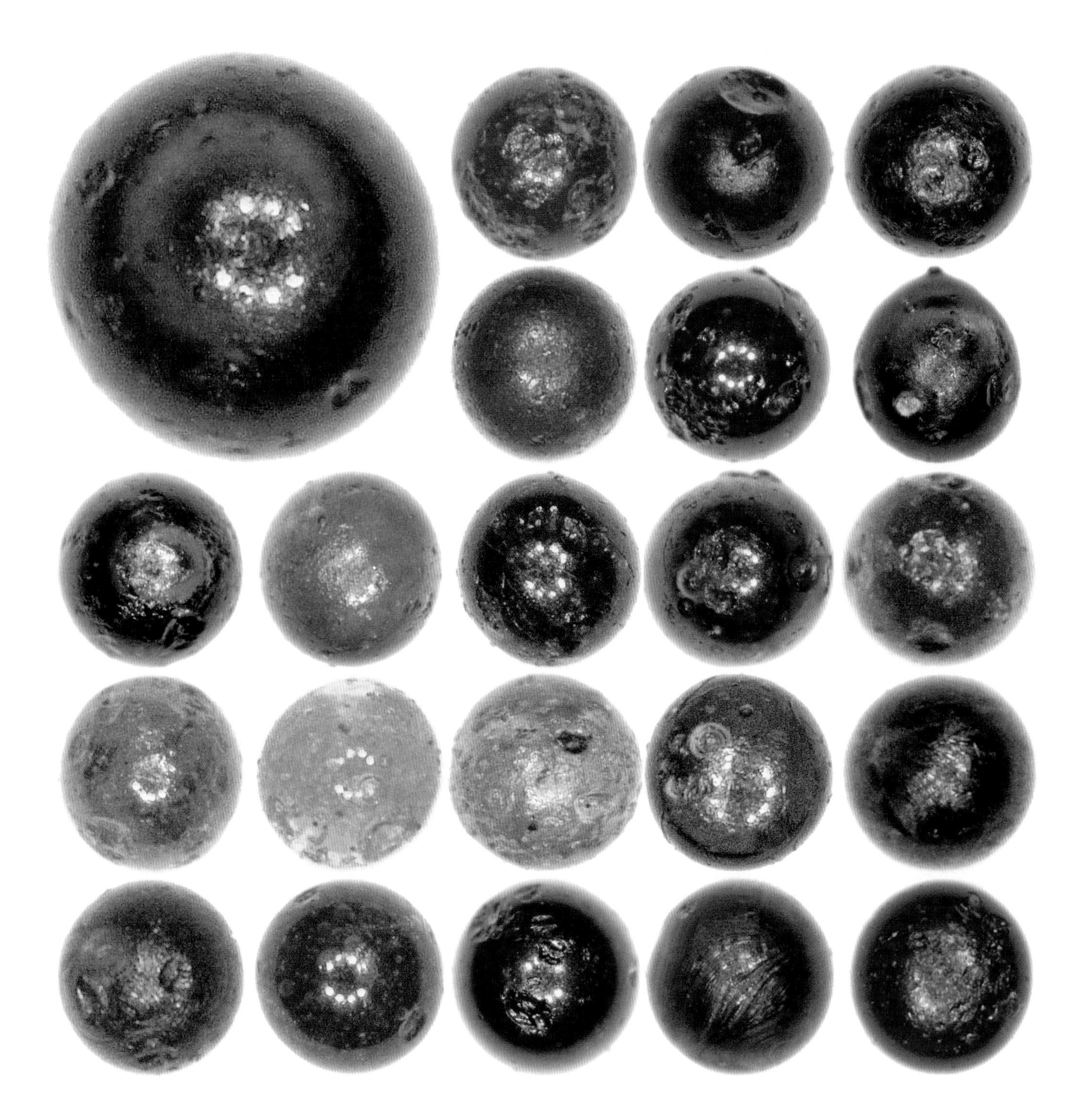

A Case Study of Anthropogenic Spherules

During five years from 2010, two to three times per year, road dust was sampled from three pedestrian refuge islands along 300 meters (984.25 feet) of Innherredsveien, a road into Trondheim city, Norway. To begin with the samples contained an anomaly—green nonmagnetic glass spherules 0.3–1.5mm, up to 75 spherules per 2,800 grams road dust—approximately one per tablespoon. The number of spherules decreased gradually over time, until zero four years later.

Uphill from the pedestrian islands was a construction site at the exit of a new road tunnel, Strindheimtunnelen. The wheels of the huge construction machines carried with them a substantial amount of fine-grained debris, clogged on the tires. This was washed down the road by rain and melting snow and accumulated around the refuge islands where the spherules were found. The origin of these unusual spherules is still not clear, but tunnel drilling, power tools, explosives, or tunnel insulation are some possibilities.

The spherules are reminiscent of the mineral wool spherules on page 99, but have a much more varied morphology. They exhibit flow banding or schlieren, percussion marks, and contain vesicles and sometimes metal beads. Together with the glass spherules a few nonmagnetic native silicon spherules were found, plus carbon cinder (see page 113) and road dust garnets (pages 147–149). Today the tunnel is open, the construction work is completed and there are no longer glass spherules like these in the road dust.

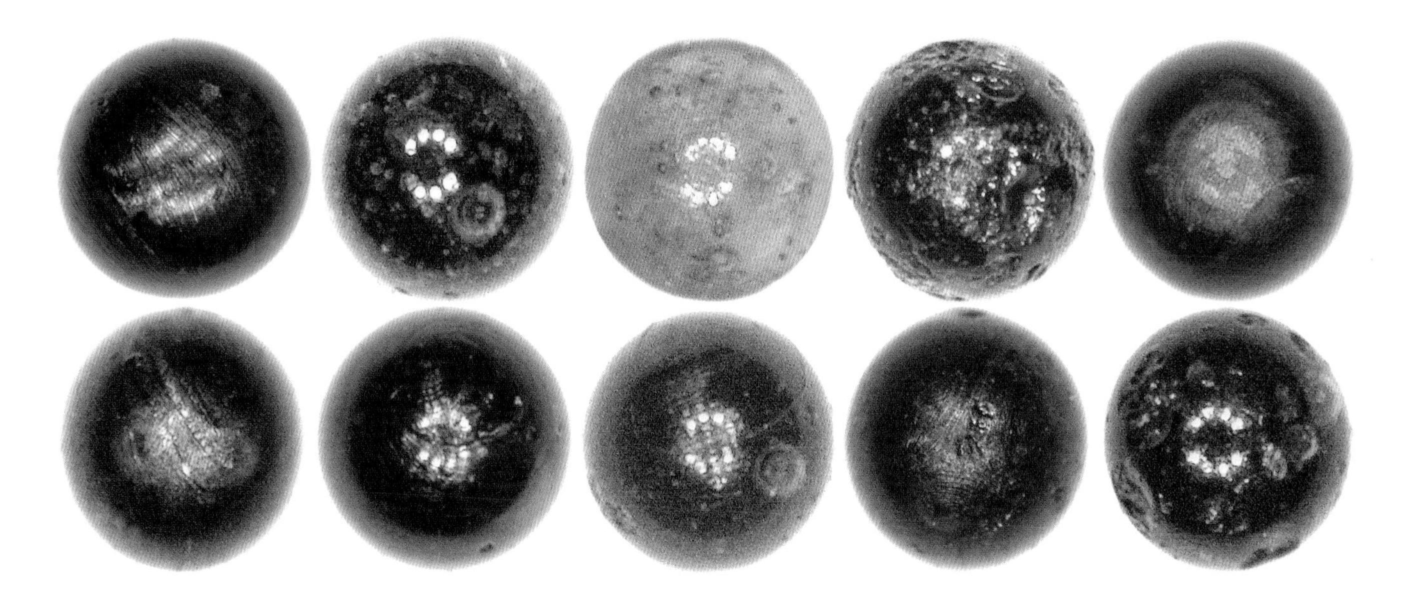

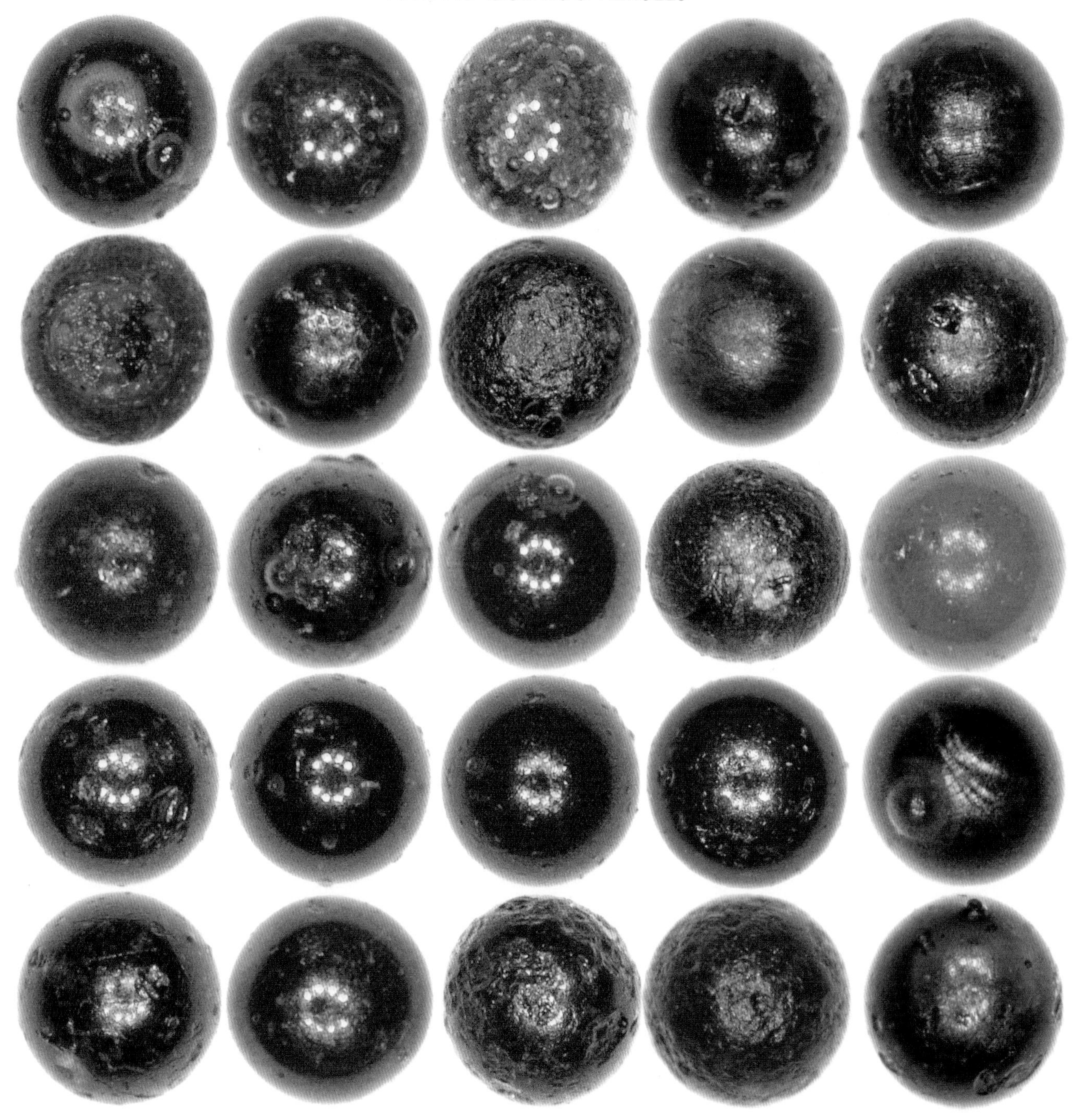

Fireworks

Fireworks and flares are chemistry at its most spectacular, but each pyrotechnic spark can create a spherule, and in certain areas these are found abundantly. Various exotic metal salts (mainly carbonates and chlorides) are used for colorization: strontium/lithium (red), calcium (orange), sodium (yellow), barium (green), aluminum/titanium/magnesium (silver/white), copper (blue), and so on. Flares are often based on strontium nitrate, potassium nitrate, or potassium perchlorate.

The majority of the spherules from fireworks can be identified as such under the microscope, either from the vivid colors, from the morphological structures, or from the characteristic metal nugget—the "eye." In most cases a chemical EDS analysis will remove any doubt.

There are, however, some problematic spherules suspected to be created by fireworks. Some seem to have a near chondritic chemistry plus barium, others look like turtlebacks (see page 27). Apparently the sparks from fireworks can go through fire and friction enough to create spherules reminiscent of micrometeorites. They will, however, lack the structure of micrometeoritic Christmas trees. This is one of the areas in need of more research in order to improve the level of precision in separating micrometeorites in populated areas.

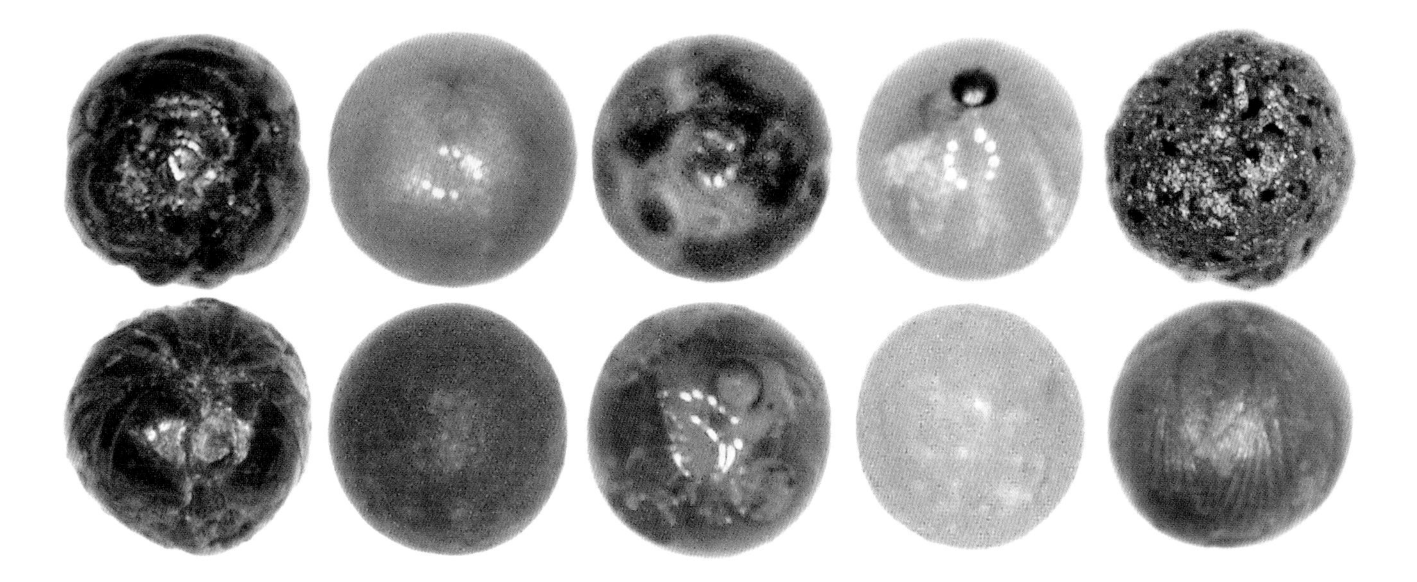

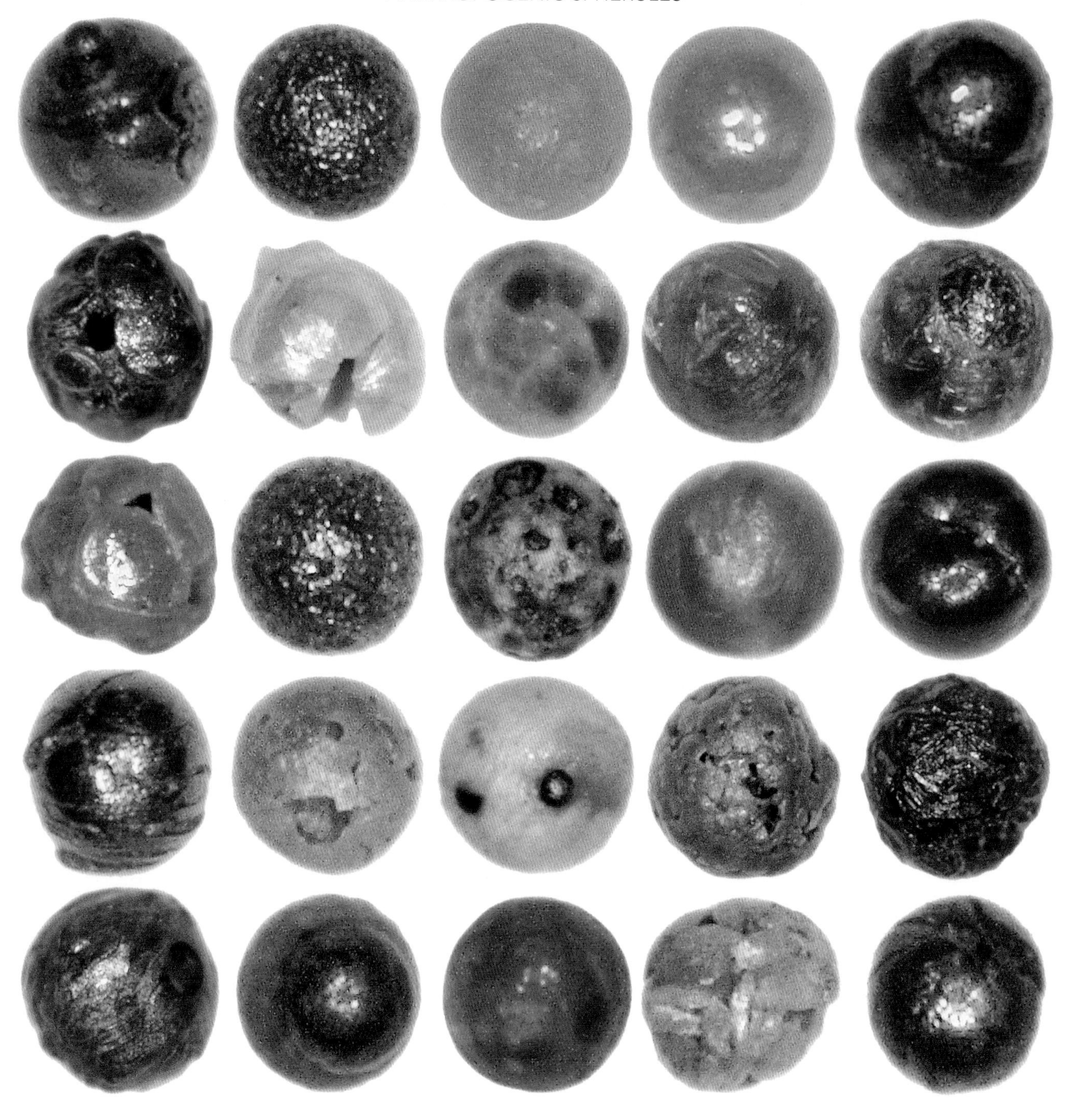

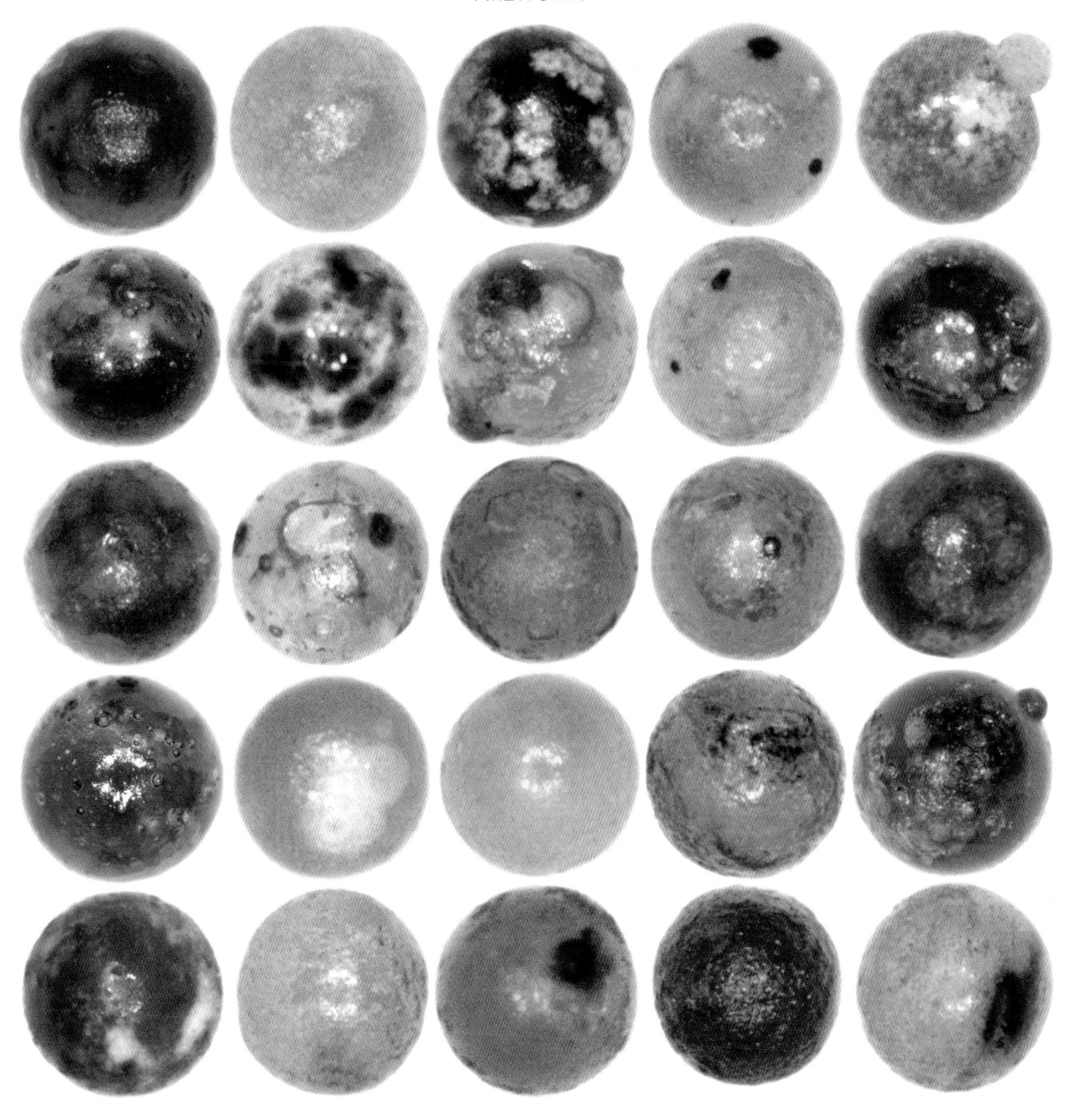

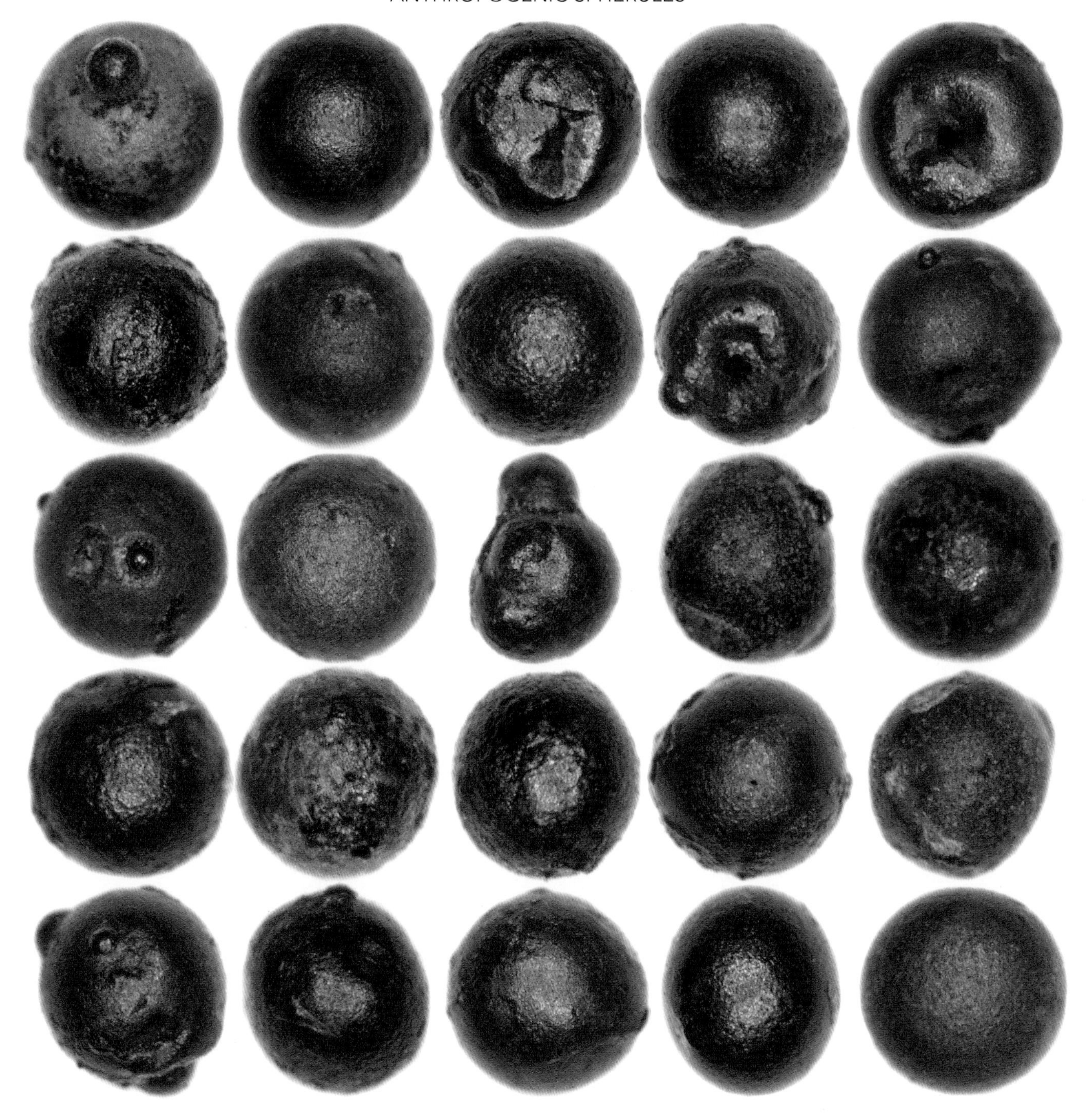

Black Magnetic Spherules

There are many references to black magnetic spherules (BMS) in older literature about micrometeorites. The term was often applied for MMs in general, but today it is no longer in use. Nevertheless, in populated areas we encounter a variety of spherules that best can be described as BMS.

The BMS are fairly common and occur abundantly locally, which in itself is an indication of terrestrial origin. The main sources for BMS are asphalt concrete, roofing shingles, and asphalt glue, which also may contain spherulitic additives like aluminium silicates or iron. The morphology goes from shiny black glass (vitreous/subvitreous) to dull gray, via metallic/submetallic, with a variety of surface structures, and from perfect spheres to composite forms. Inside, the black magnetic spherules are most often homogenous black, sometimes with a small central vesicle, and with a radiating structure.

The black carbon (fly ash) is produced by incomplete carbon combustion, both natural and industrial, and these aerosol particles are found worldwide. They are easily distinguished from the types of BMS in this album by the difference in size: black carbon has an average diameter of 0.0025 mm (2.5 µm), which is less than 1⁄100 of an average micrometeorite.

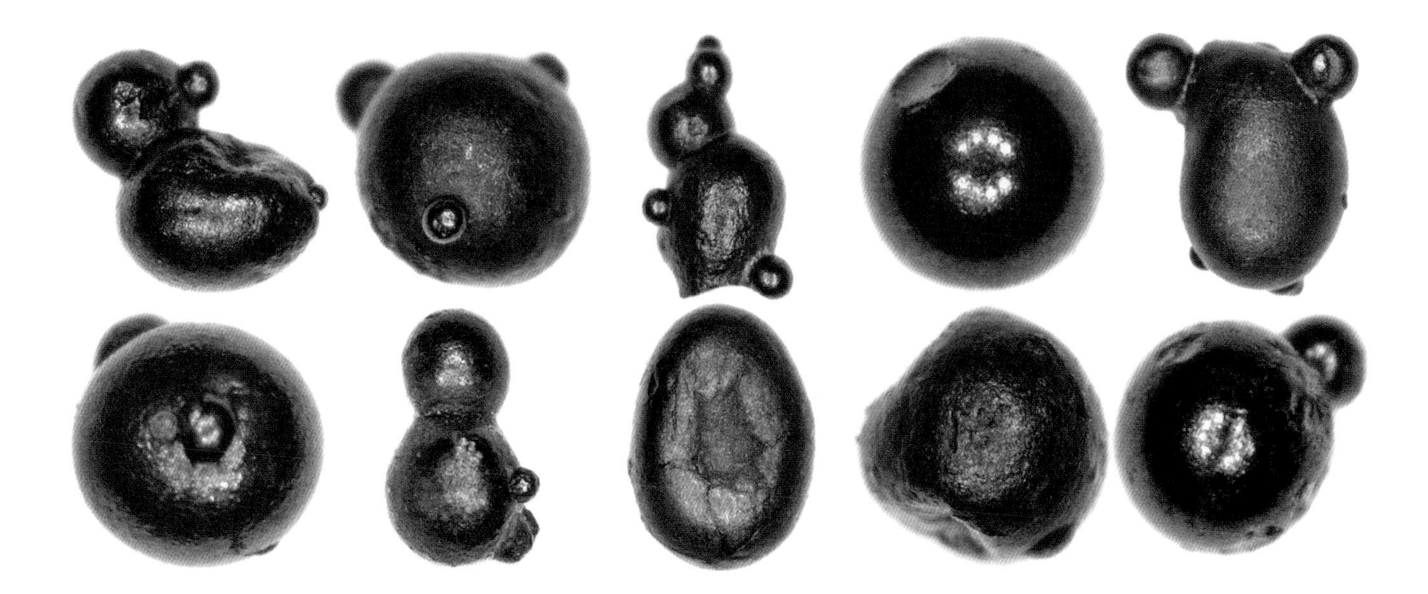

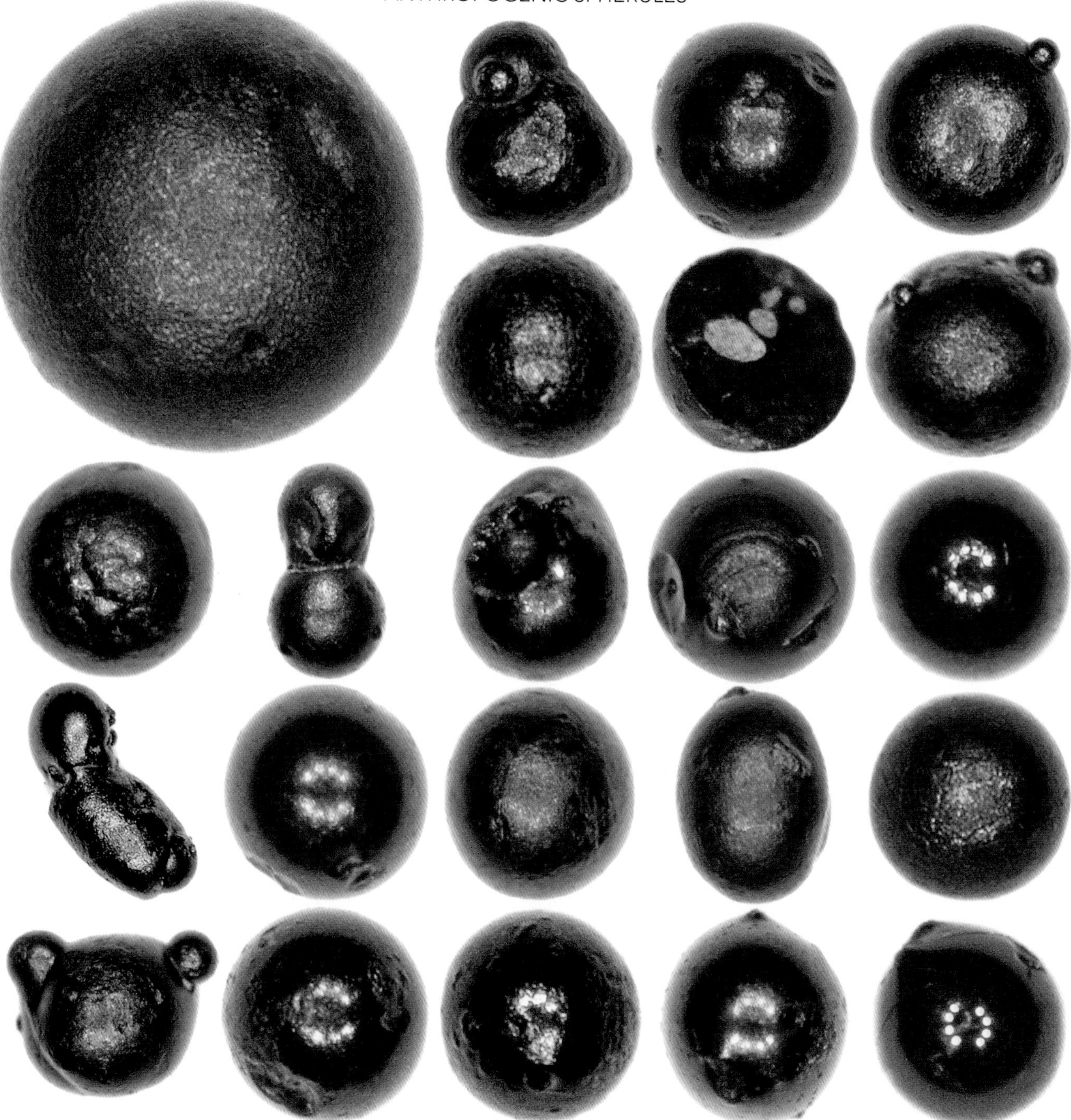

Roof Tiles and Shingles

When searching for micrometeorites on roofs one may encounter various erosion products of the roof itself. Ceramic tiles on a pitched roof are traditionally made from terracotta or concrete and often covered with fine sand in pigmented iron rich glue. Over time these magnetic particles (0.2–1.0 mm) will erode and sometimes fill the rainwater gutter. For erosion products from asphalt (bitumen) shingles, see pages 106–109.

A roof composed of slate shingle will erode to mainly nonmagnetic mineral particles, and provides an excellent hunting ground for micrometeorites.

Metallic Carbon Cinder

These strange objects can be found at many locations around the world, but sparsely. They are apparently some sort of metallic cinder, at times with a silky luster, size ~0.3–1.2 mm. They are not magnetic, and a chemical EDS analysis reveals that they are practically pure carbon.

These glassified carbon particles have been found both in contemporary sediments as well as in 32,000-year-old clay, but the origin remains a mystery. Glass-like carbon is also reported from Younger Dryas in connection with the Clovis comet hypothesis. Furthermore, the rare carbon meteorites CI chondrites seem to be the next of kin of the micrometeorites, but their chemical spectrum is far more complex than this carbon cinder. A wild guess is fly ash—from burning of wood?

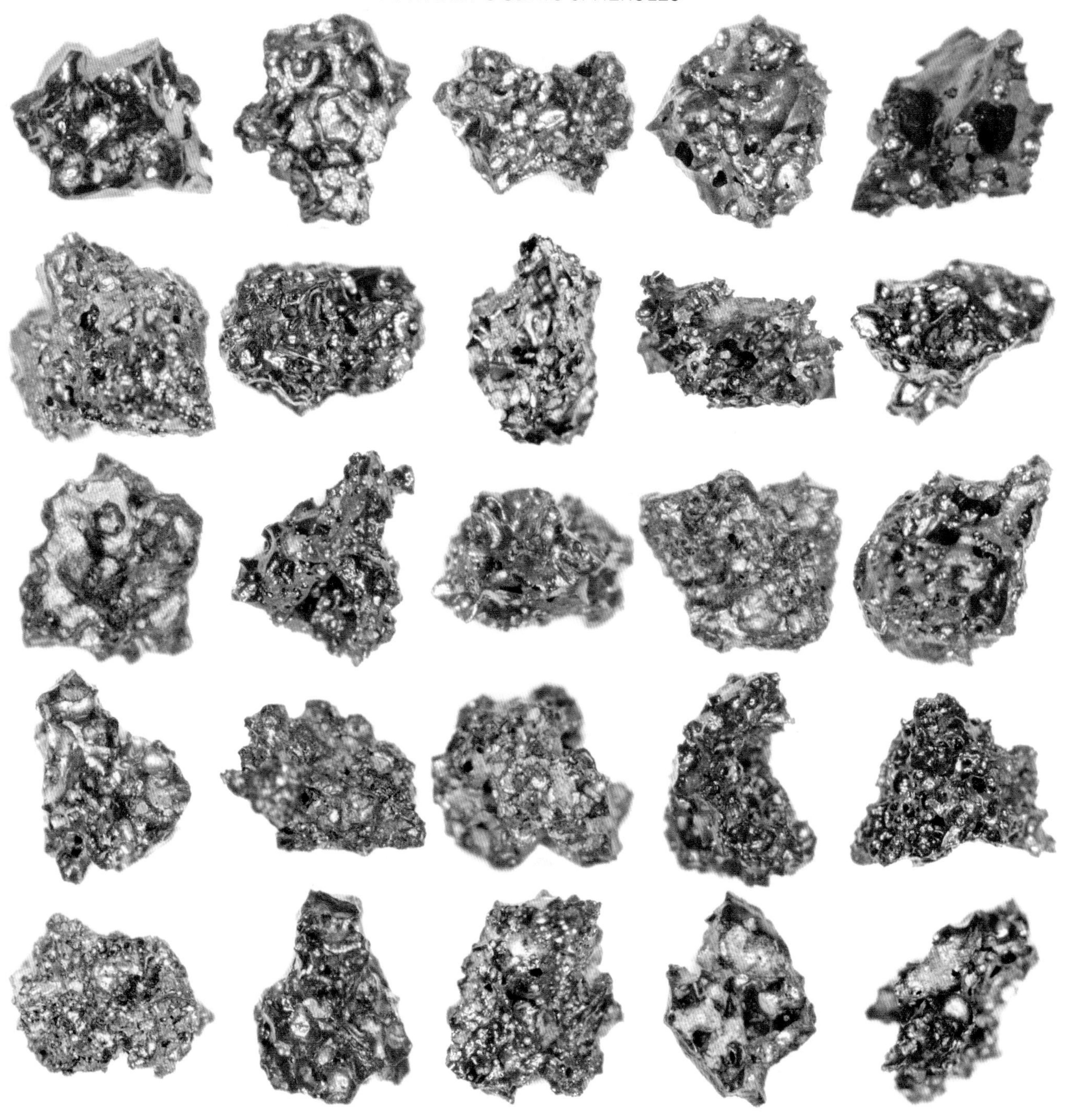

Red Scoriaceous Spherules

These spherules can be found in many places around the world, but mainly in low numbers. They are magnetic, and show an impressive variety of morphological features. The interior of the spherules is usually opaque black, glassy, and vesicular. Sometimes the vesicles are filled with white minerals or the vesicles can be open with a smooth glassy surface, occasionally with an iron oxide rim. The exterior is reddish/brown, often with white/yellow protuberances and sometimes with a partial iron oxide rim. They often have a central chimney hole where volatile elements may have escaped.

The origin of these spherules is uncertain, but clearly terrestrial—possibly fulguritic (see page 126), or products from some sort of combustion (page 96) or perhaps slag from welding. The size is ~0.1–10.0 mm, which is too large to be exhaust from any ordinary combustion engine. They occur frequently in the rain gutters of tall buildings, which strengthens the suspicion that they may be fulgurites.

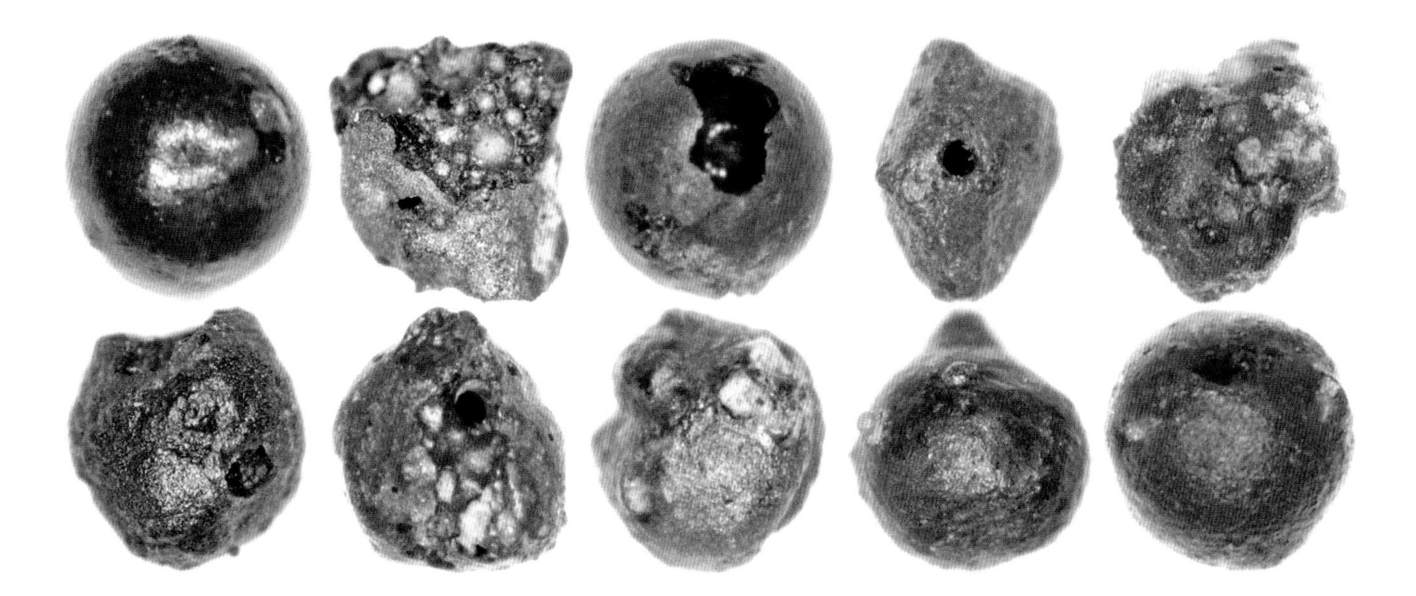

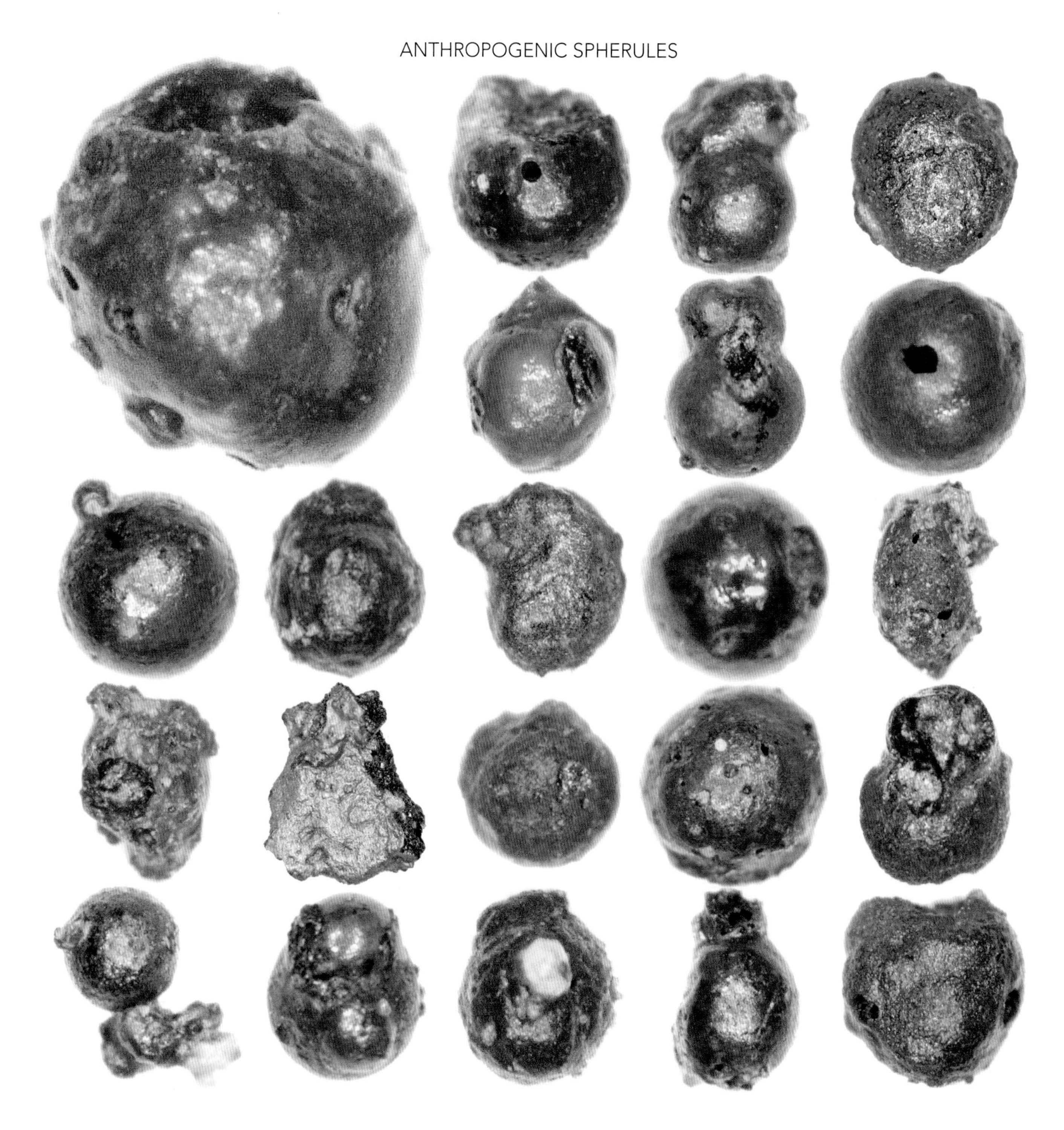

RED SCORIACEOUS SPHERULES

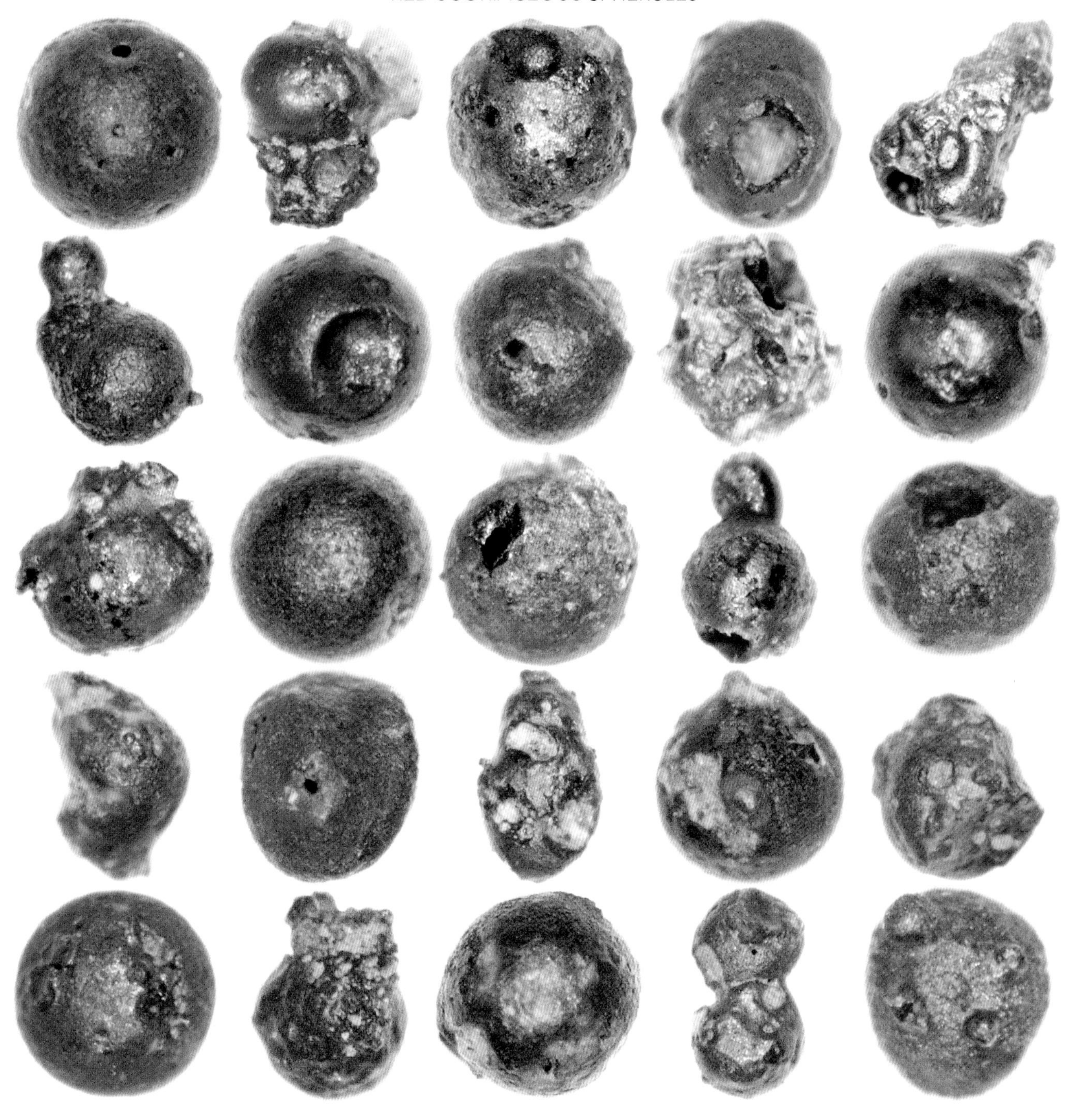

IV TERRESTRIAL OBJECTS

Rounded Mineral Grains

One of the first types of rounded objects we are likely to encounter in the search for cosmic spherules is sand—mineral grains more or less rounded by erosion. Huge amounts of sand are transported around the world by the wind, and the global mineral dust emissions are estimated at 1,000–5,000 millions of tons per year, of which 60–200 million tons originate from the Sahara desert alone. A substantial amount of this is transported across the Atlantic into the Caribbean and Florida.

Consequently rounded mineral grains will occur in the most unexpected places. A morphological examination under a microscope is usually enough to determine whether the candidate has any of the characteristic structures of a micrometeorite.

Magnetite

Searching for micrometeorites with a magnet is the King's Road to success. Wash/dry and fractionate the dust samples to 0.1–0.4 mm, and then extract the magnetic particles. This way the possible micrometeorite specimens will be enriched. Improve the odds by sampling from a large surface area that has accumulated particles from above over a long period of time, preferably with low accretion of terrestrial mineral particles.

There is, however, one mineral that will occur practically everywhere; it is the most magnetic of all the naturally occurring minerals on Earth—magnetite. When sampling for micrometeorites in road dust, beach sand, deserts, mountains, or even on roofs, the magnetite particles are abundant and will get caught by the magnet, like the crystals on these pages.

Fulgurites

In nature there are three processes that can melt rock: volcanism, meteorite impacts, and lightning, and when we search for cosmic spherules we look for melted droplets. These molten micrometeorites distinguish themselves from other mineral grains with their ablated spheroid form and characteristic structures. But if we look for molten mineral grains we will also encounter other types—like the fulgurites.

Fulgurites are well known as "lightning tubes" (glass) from molten desert sand. On rare occasions a lightning bolt can strike directly on rock, melt a hole and splash droplets, or exogenic fulgurites, can be formed. A recently discovered third type is created when the lightning strikes within the dust plume of an erupting volcano—exogenic volano-fulgurites.

There is a fourth type, exogenic phyto-fulgurites—possibly the most common of the four, but still not properly described. The spherules on these pages were found at a place where the lightning had struck a tree and gone to ground via humus and soil. The EDS analysis shows a surprisingly homogenous chemical spectrum despite the wide color range: aluminum silicates (sometimes with an iron oxide rim), and hardly any trace of carbon. This will vary according to the content of the target soil. Approximately one third of these spherules are magnetic, mainly those with a visible metallic (iron oxide) rim. The spherules are 0.2–6.0 mm, found at Ann Arbor and Ypsilanti, Michigan, US.

On a global scale there are approximately 100 lightning strikes per second, and obviously these resistant silica spherules will not easily dissolve and disappear, so this type of spherule can be expected to be found everywhere (see also page 115).

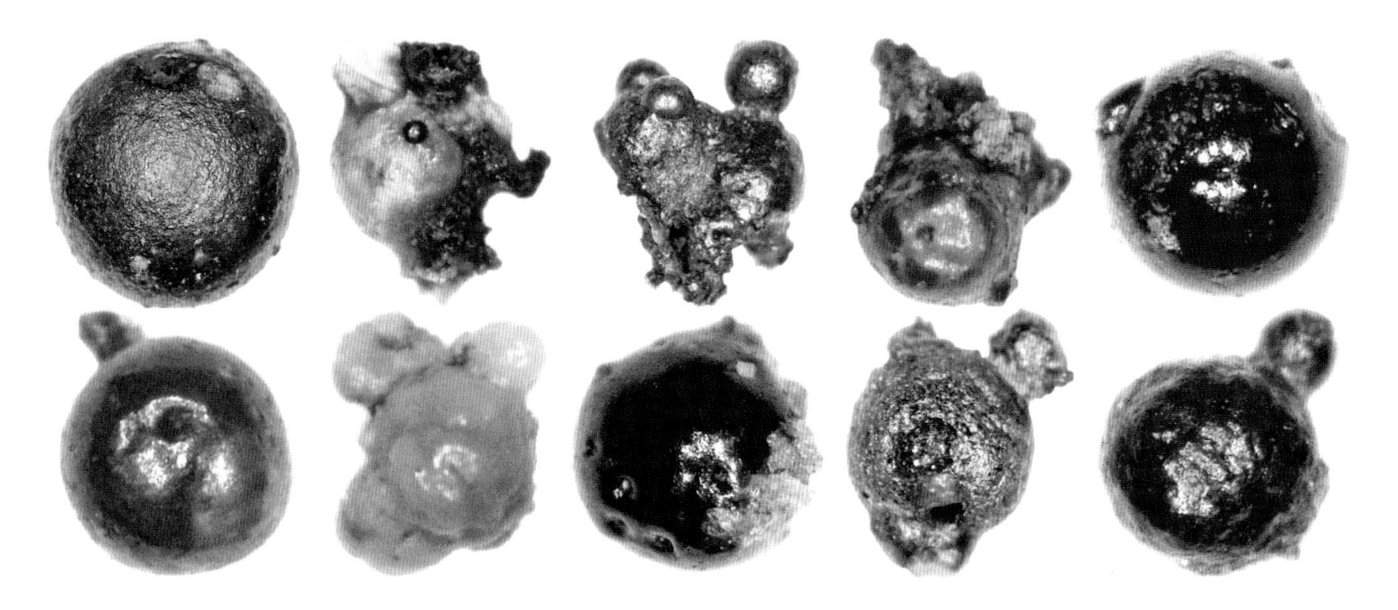

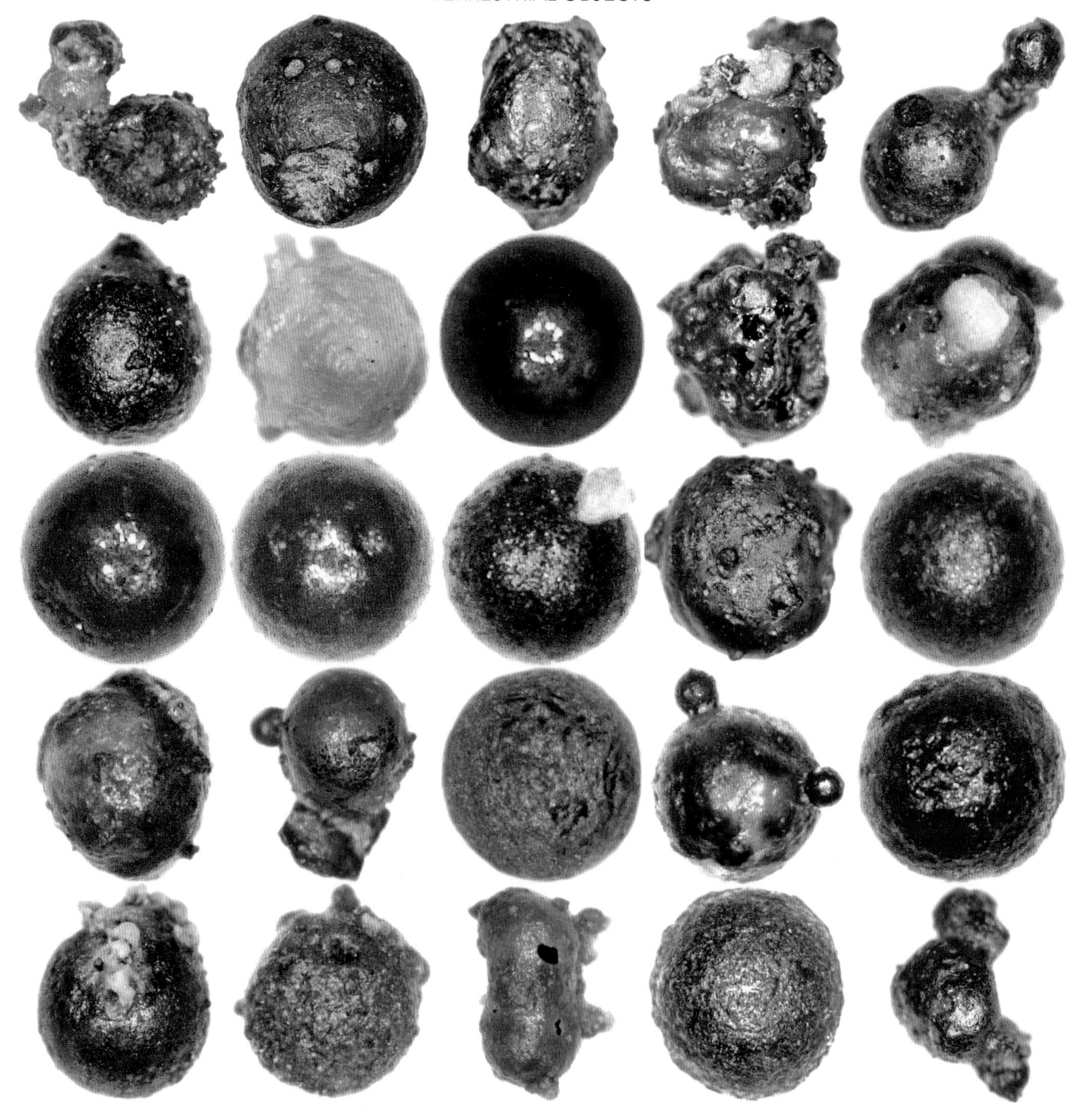

Organic Confusion

An interesting stage in the hunt for micrometeorites comes after the field search: visual examination of the dust samples under the microscope. I start with a binocular microscope and pick out promising candidates for further scrutiny under a USB microscope with stronger magnification. But first the dust sample must be washed, dried, and fractionated. For the latter I use two sieves; 1.5 and 0.4 mm.

The greater than 1.5 mm fraction is mainly to get rid of these particles so the medium fraction (1.5–0.4 mm) can be examined. Here the particles are large enough to clearly see the surface structures, but giant micrometeorites this size are rare. In the fine, less than 0.4 mm fraction the particles are numerous, and patience is of the essence. This is where we may find the extraterrestrial objects in the proverbial search for the needle in the haystack. According to Michel Maurette the particle size distribution of the micrometeorites have a peak in size between 0.2 and 0.4 mm.

Consequently, if over time you search for MMs in dust samples under a microscope, you will encounter countless terrestrial particles of all sorts. Sometimes a possible candidate or an unusual object is put aside for thorough inspection, but further investigation may reveal an organic origin: plant seeds, snail shells, insect debris, micro fossils, feces, fungi, sclerotia, etc. Nature is full of beauty, and here are photos of particles that have caused both organic confusion and joy in the hunt for the extraterrestrial.

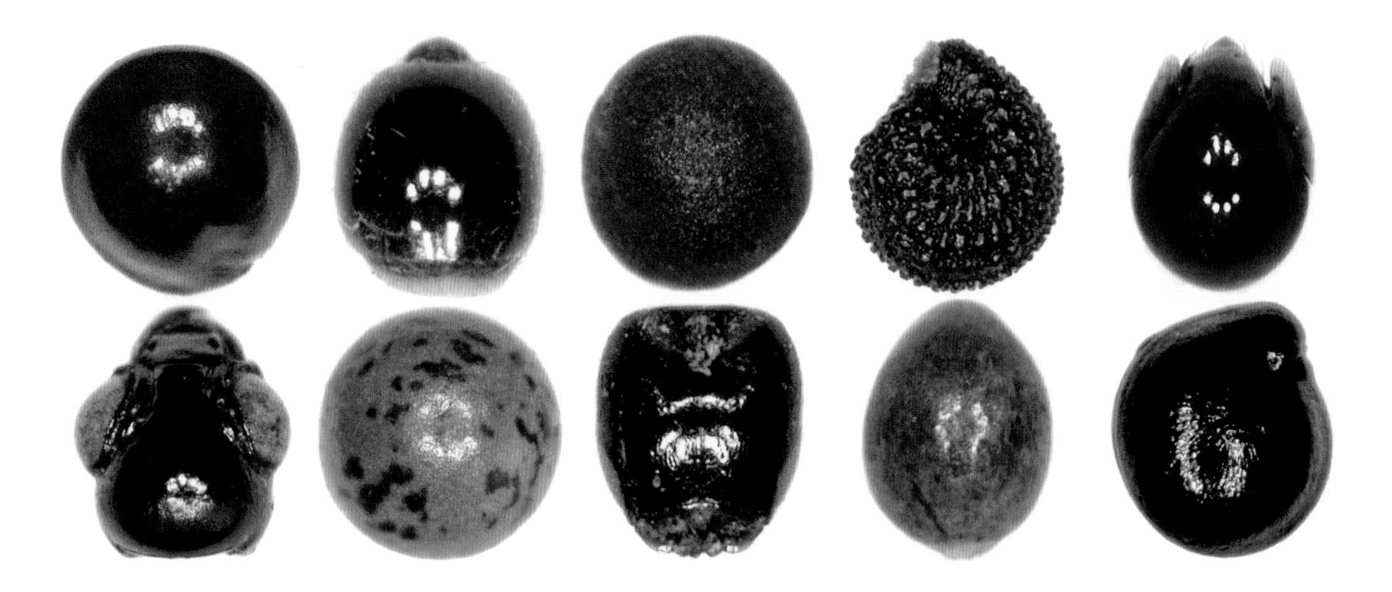

PHOTO: LAURENCE GARVIE

METEOR CRATER IMPACTITE SPHERULES

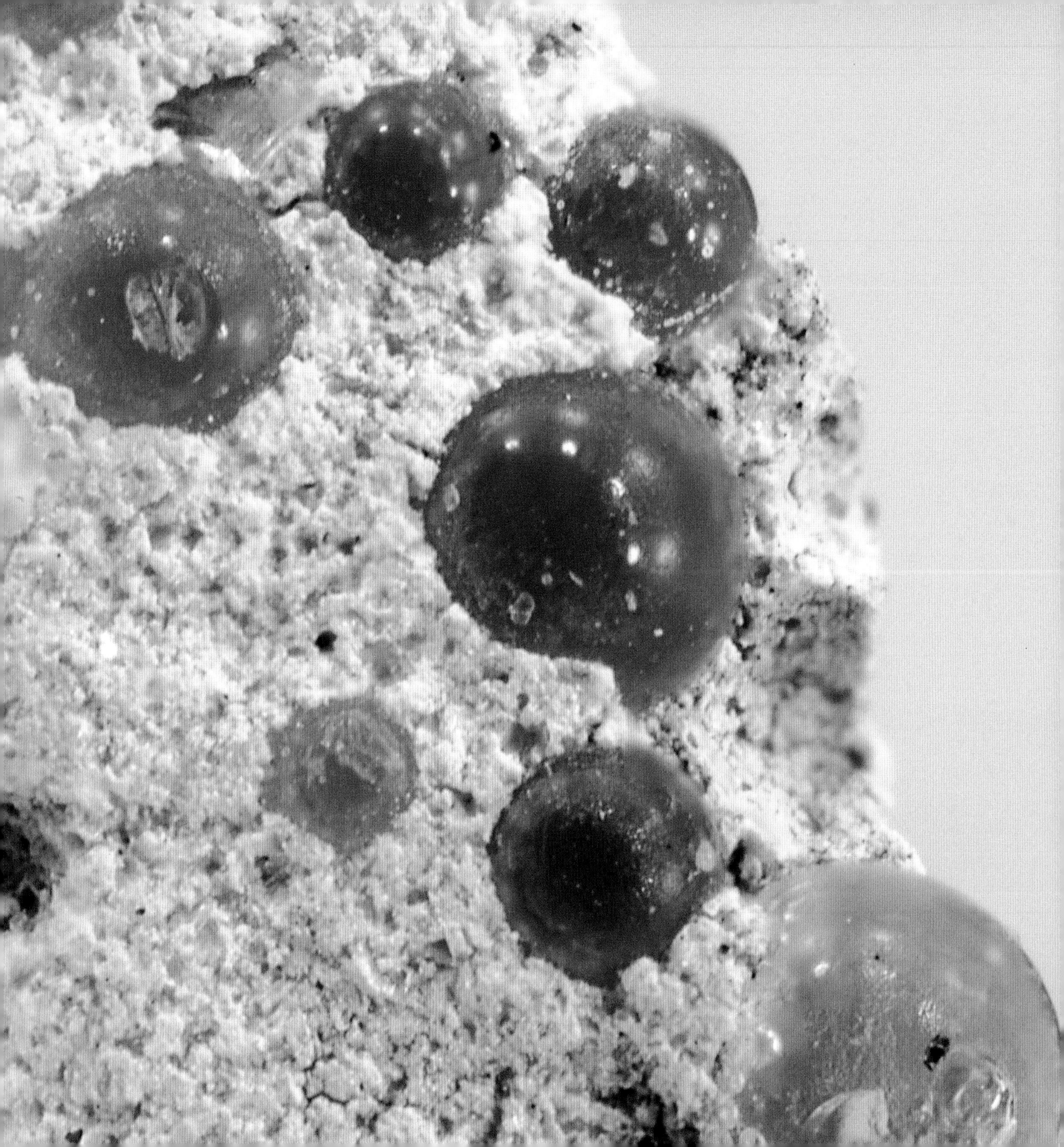

Microtektites and Microkrystites

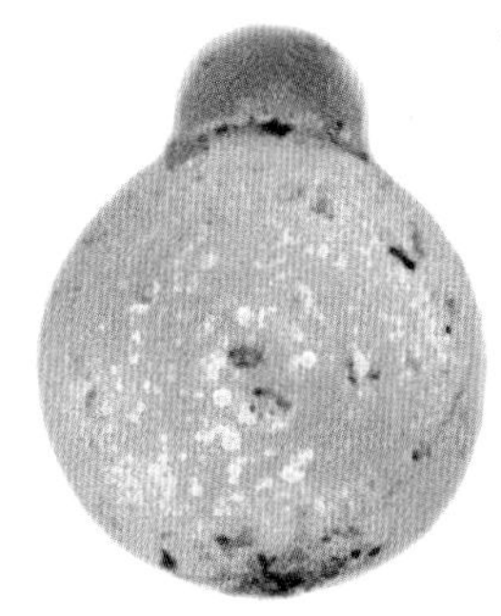

Microtektites and microkrystites are submillimeter-sized spherules formed by the large asteroid impacts on Earth. They predominantly comprise melted and vaporised terrestrial target rock. Microtektites, which are wholly glassy by definition, can be derived from individual melt droplets, as ablated macrotektite material and as vapor condensates, which form a distinct group. Microkrystites are thought to be vapor condensates, and differ from microtektites in that they are part glassy and partly crystalline. Proximal impactite splash droplets are likely to have a degree of crystallinity due to a more basic composition derived from a greater degree of meteoritic contamination of the target rock.

Microtektites are found in defined strata in three of the five established tektite strewnfields on Earth. Spherules can also be found in connection with other smaller-sized craters, like the rare S-type spherules from Meteor Crater, Arizona, on page 132. The most famous (altered) microtektites are those found at the iridium-enriched Cretaceous-Paleogene boundary, originating from the great Chicxulub Crater at Yucatán, Mexico. The spherules on this page are from this event 66 million years ago. The one at the top was found in Saskatchewan, Canada, and the black ones in Hell Creek, Hardin, South Dakota, USA.

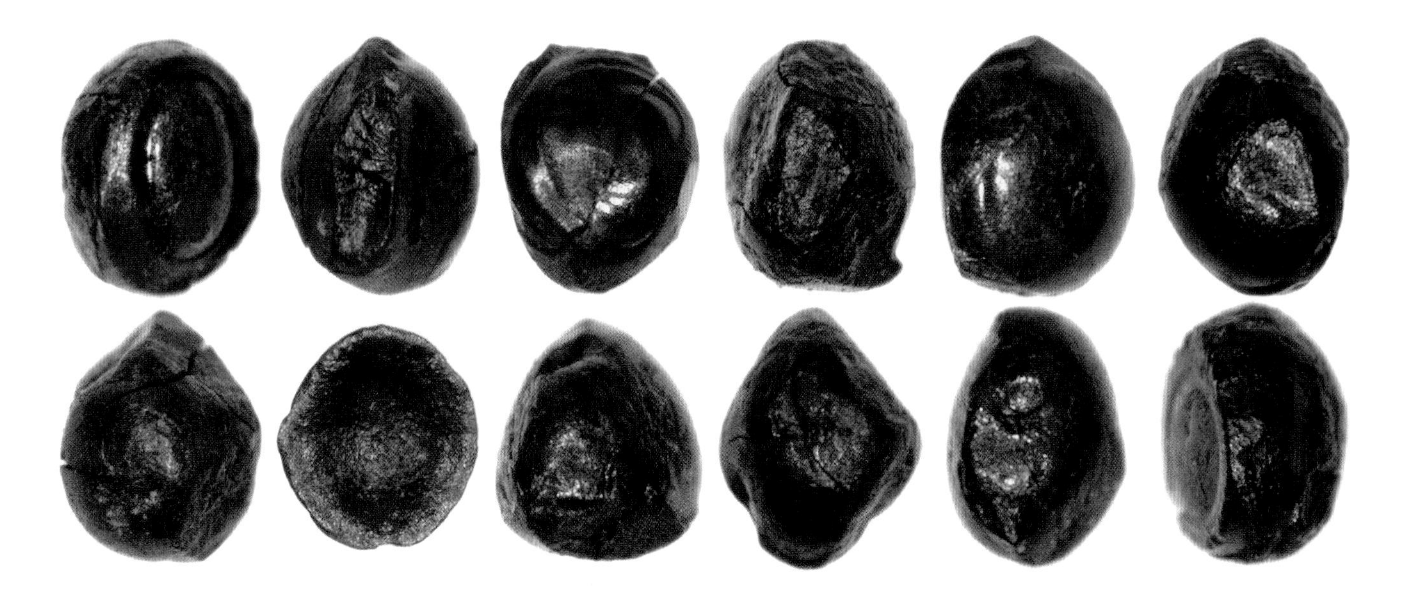

Lonar Crater Spherules

In the Buldana district, Maharashtra, India, there is a crater 1.88 km in diameter, created by a meteorite impact estimated at more than half a million years ago. Scattered around the well preserved crater are impactite droplets that differ from most other spherules on Earth, but assumed to be far more common on the Moon and Mars—basaltic impactite melt droplets. Compared with the chemical composition of the surrounding basalt, the spherules are enriched in chromium, cobalt and nickel, interpreted as remnants of a chondritic meteorite impactor.

The spherules on these two pages are collected at the Lonar crater rim. They are slightly magnetic, and between 0.5–14 mm.

Darwin Glass

Darwin glass impactite droplets are found in a 410 km^2 strewn field south of Queenstown, West Coast Range of Tasmania. They originate from the Darwin Crater, a meteorite crater 1.2 km in diameter created by a 20–50 meter meteorite impact approximately 816,000 years ago. These 4–8 mm melt droplets were found 4–5 km from the crater.

Whereas micrometeorites are evenly distributed globally, microtektites and impactites like the Darwin glass occur only in the local strewn field, and only in one defined stratum.

The white/green impactite spherules are mainly composed of melted local metamorphic rocks, but the black/green ones contain less silica and more magnesium, iron, chromium, nickel and cobalt—possibly components of the extraterrestrial material from the meteorite.

Volchovites—A Russian Mystery

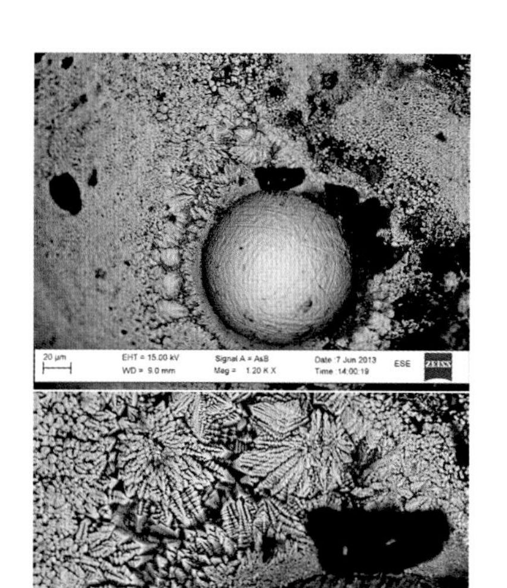

The amazing volchovites are tektite-like glasses/microcrystites 0.1–3.0 mm of mafic and ultramafic composition found in the glaciofluvial drift along the Volkhov river, near St. Petersburg, Russia. They were discovered and described by Gennady Skublov as crypto-volcanic glass. In a recent study the volchovites are subdivided into four undergroups according to the trace elements found in the spherules.

The two SEM images (left) show the microcrystite structure of a volchovite with relative large crystals around an iron oxide microspherule.

The volchovites are reported to occur together with particles of quenched glass, cinder (see page 113) and rounded fragments of rocks. Some of the spherules contain small metal beads with various contents: titanium, iron, gold, and copper—like the one on the page 138, frozen in time in the very last moment before escaping.

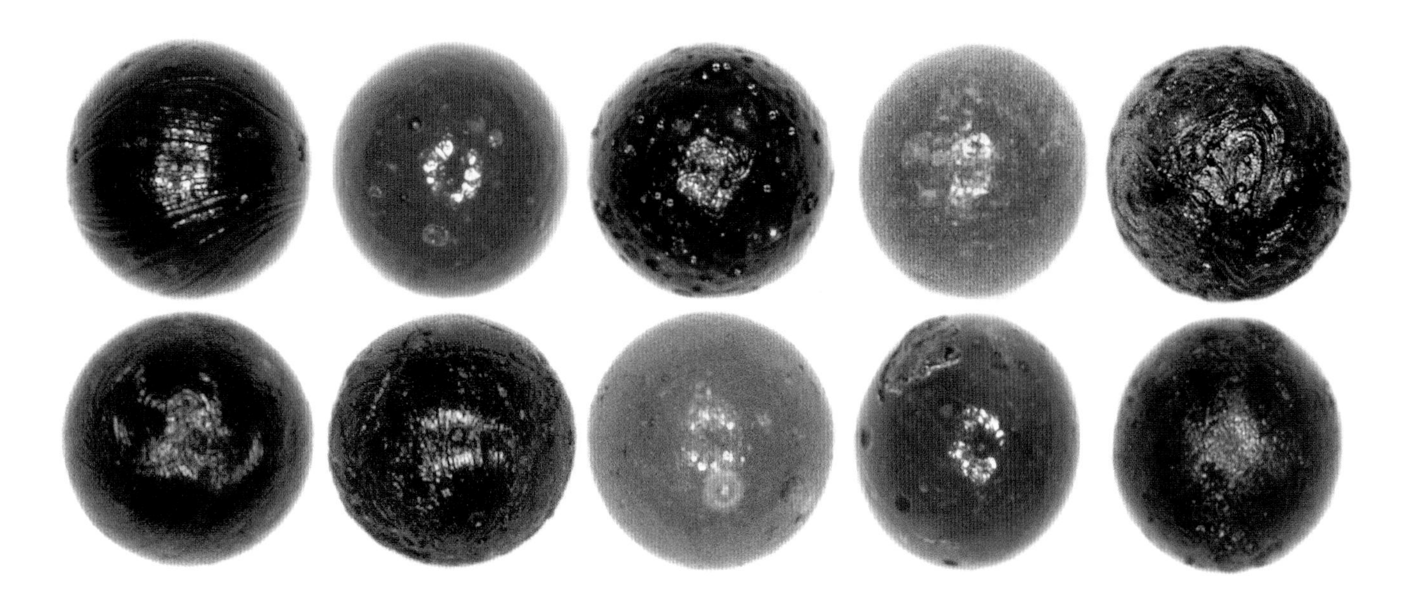

VOLCHOVITES

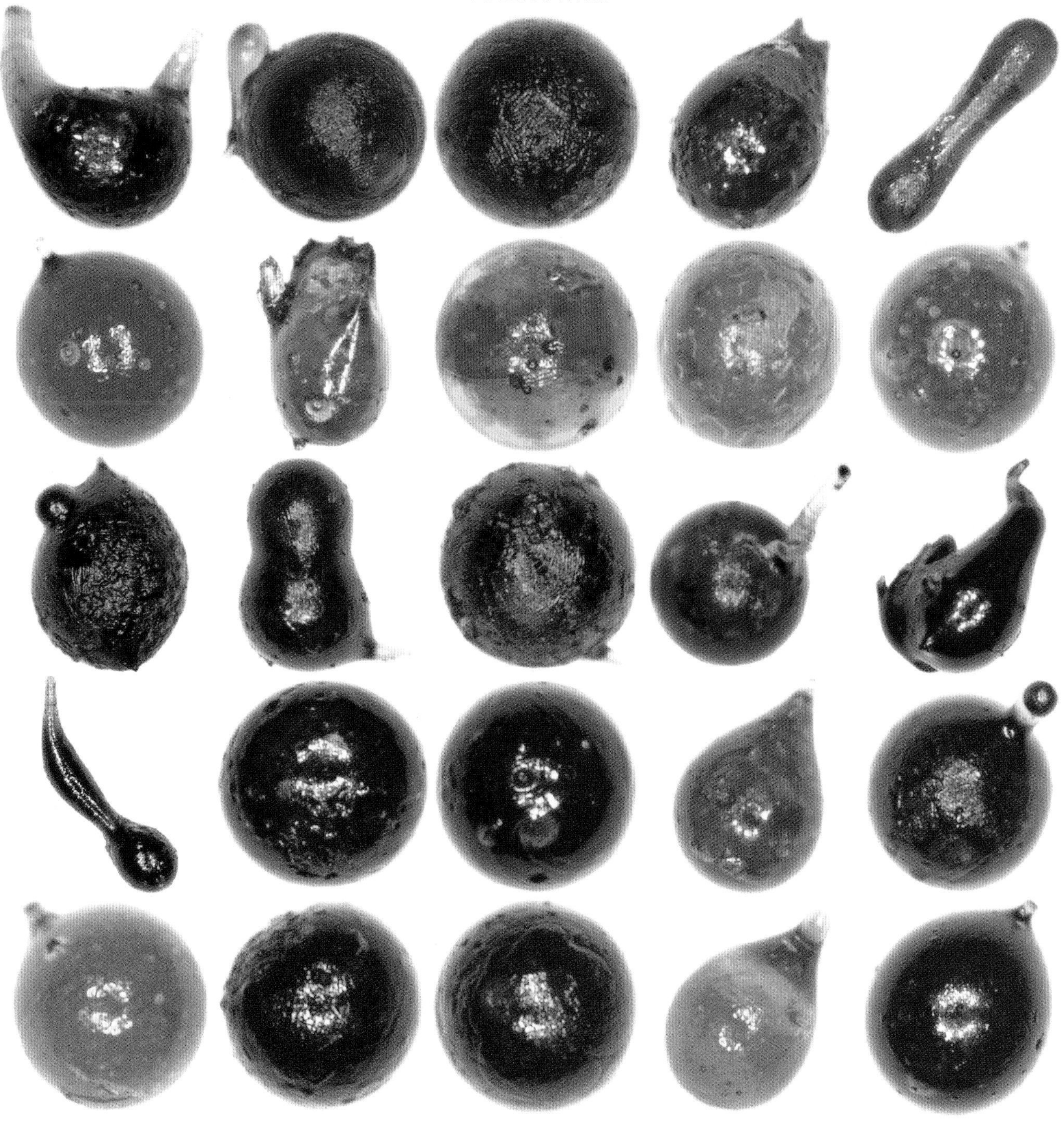

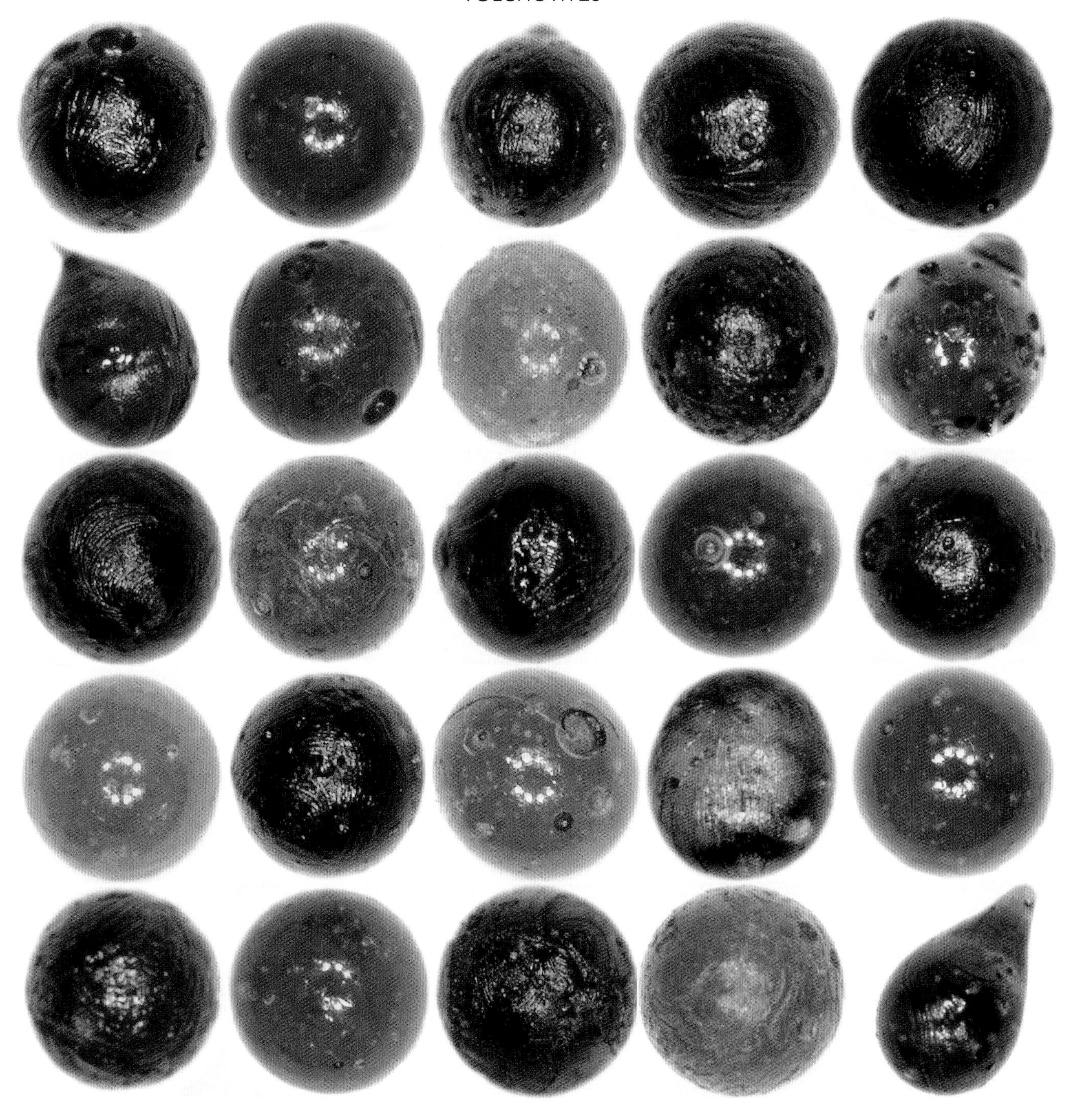

Iberulites

The Iberian Peninsula is regularly sandblasted by particles from Africa. In between the sand grains (see page 123) are pale spherules of a different kind. They are white to sand colored, often with a vertical axis, and sometimes with a characteristic vortex. Some have a smooth white surface, other seem to be a ball of sand glued together with carbonates (react to acid). Contrary to the regular sand grains, they are easily crushed, which reveal a coarse grained core (mineral particles in the 1–2μm fraction), with a thick rind of fine-grained white clay minerals.

Iberulites develop in the troposphere before they fall to Earth's surface. They are linked to the evolution of high-dust air plumes which originate in Saharan dust storms, and are transported over the Iberian Peninsula and often across the North Atlantic Ocean. One may assume similar spherules are created also in the other grand deserts of the world.

The 35 spherules (~0.3–0.9 mm) on these two pages are found along the Costa del Sol, Spain.

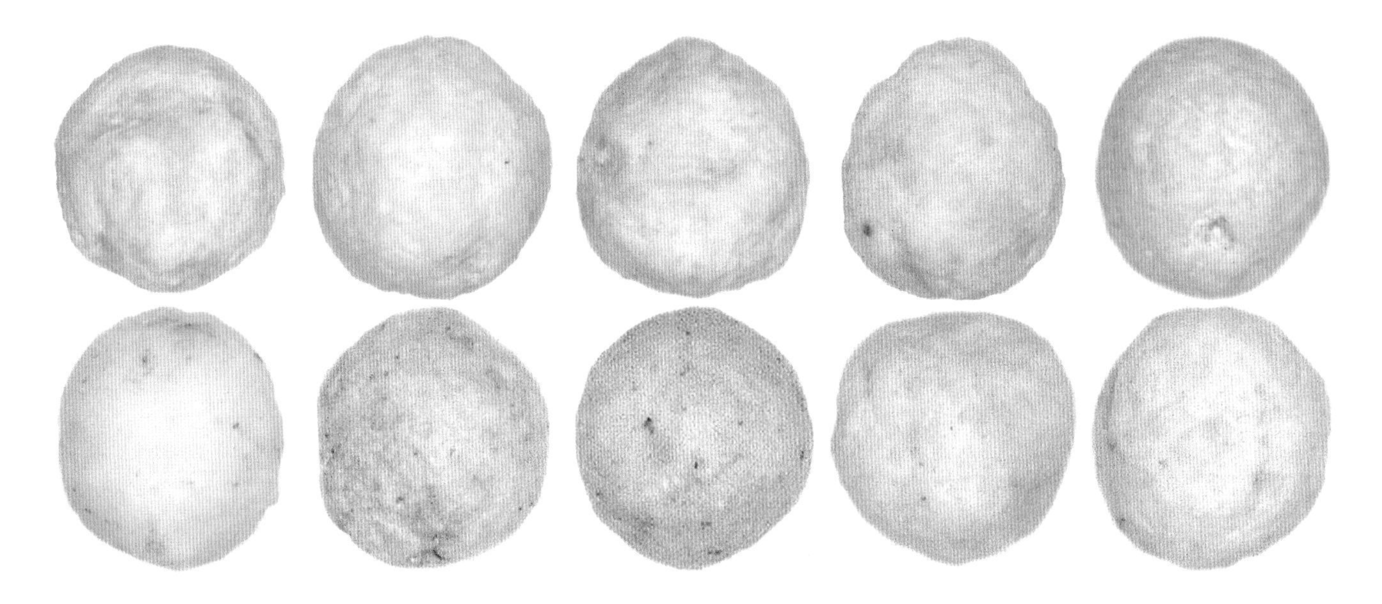

IBERULITES

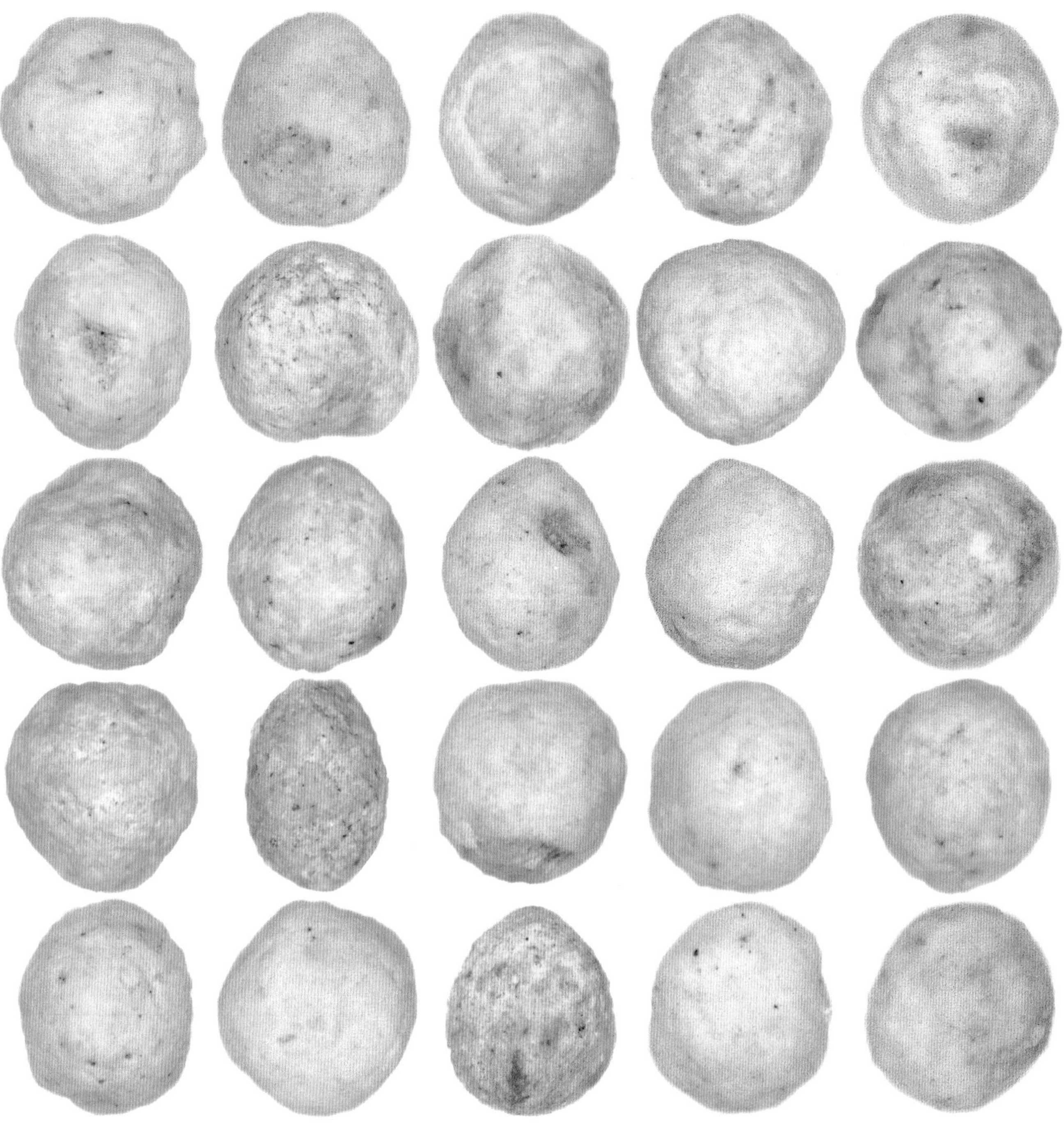

Ooids and Pisoids

Micrometeorites are rare, so spherules found in abundance is a sign of terrestrial origin. Here is a type that can be rock-forming, and as such can hardly be of any confusion in the hunt for the extraterrestrial, but are nevertheless worth being aware of. Don't let these terrestrial imposters fool you.

An ooid consists of a nucleus (usually a mineral grain or biogenic fragment) around which concentric layers of minerals are deposited to form a spherical grain of 0.25–2 mm. Oolith is sediment consisting of ooids. They are most commonly composed of calcium carbonate (the minerals calcite or aragonite), but can be composed of phosphate, silica (chert), dolomite, or iron minerals.

The ooids are usually formed in a warm, shallow, highly agitated intertidal marine environment, though some are formed in inland lakes. The mechanism of formation starts with a small fragment acting as a "seed" (nucleus), and strong intertidal currents wash these seeds around on the seabed where they accumulate layers of chemically precipitated calcite from the supersaturated water. The accreted mineral cortex increases in sphericity with distance from the nucleus. Oolith is commonly found in large current bedding structures resembling sand dunes.

The freshwater pisoids 4.0–5.0 mm on these pages are from the Cretaceous period, about 112–97 million years ago, found near Taouz, southeast of Erfoud, in Morocco.

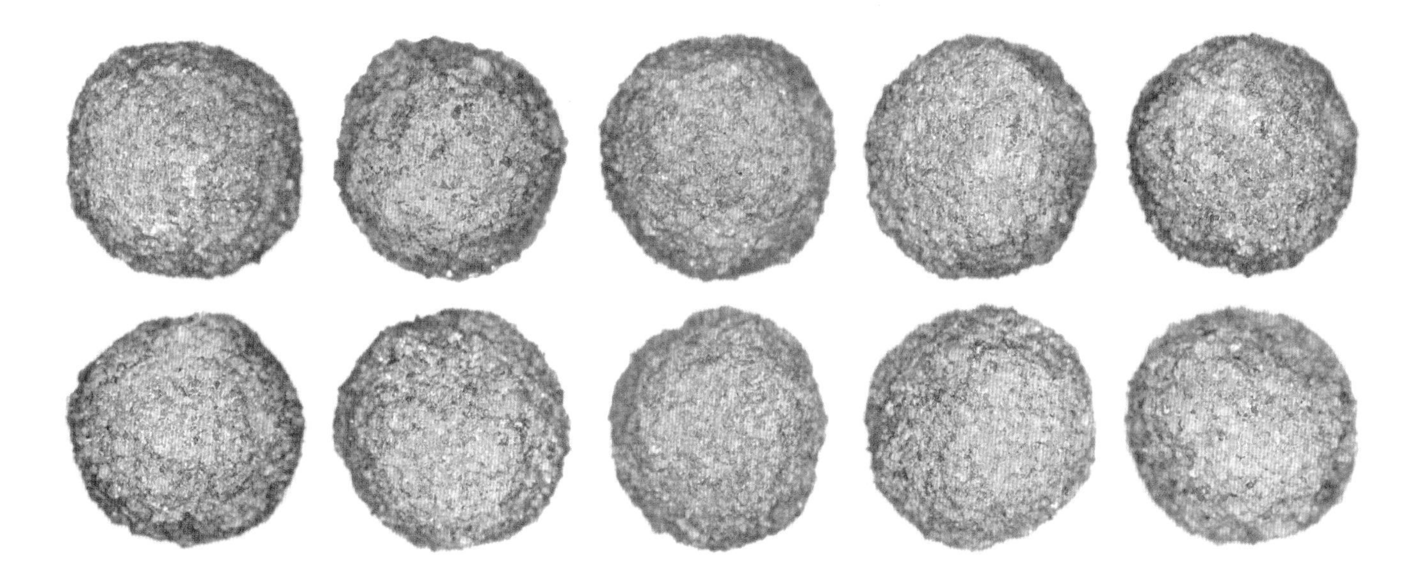

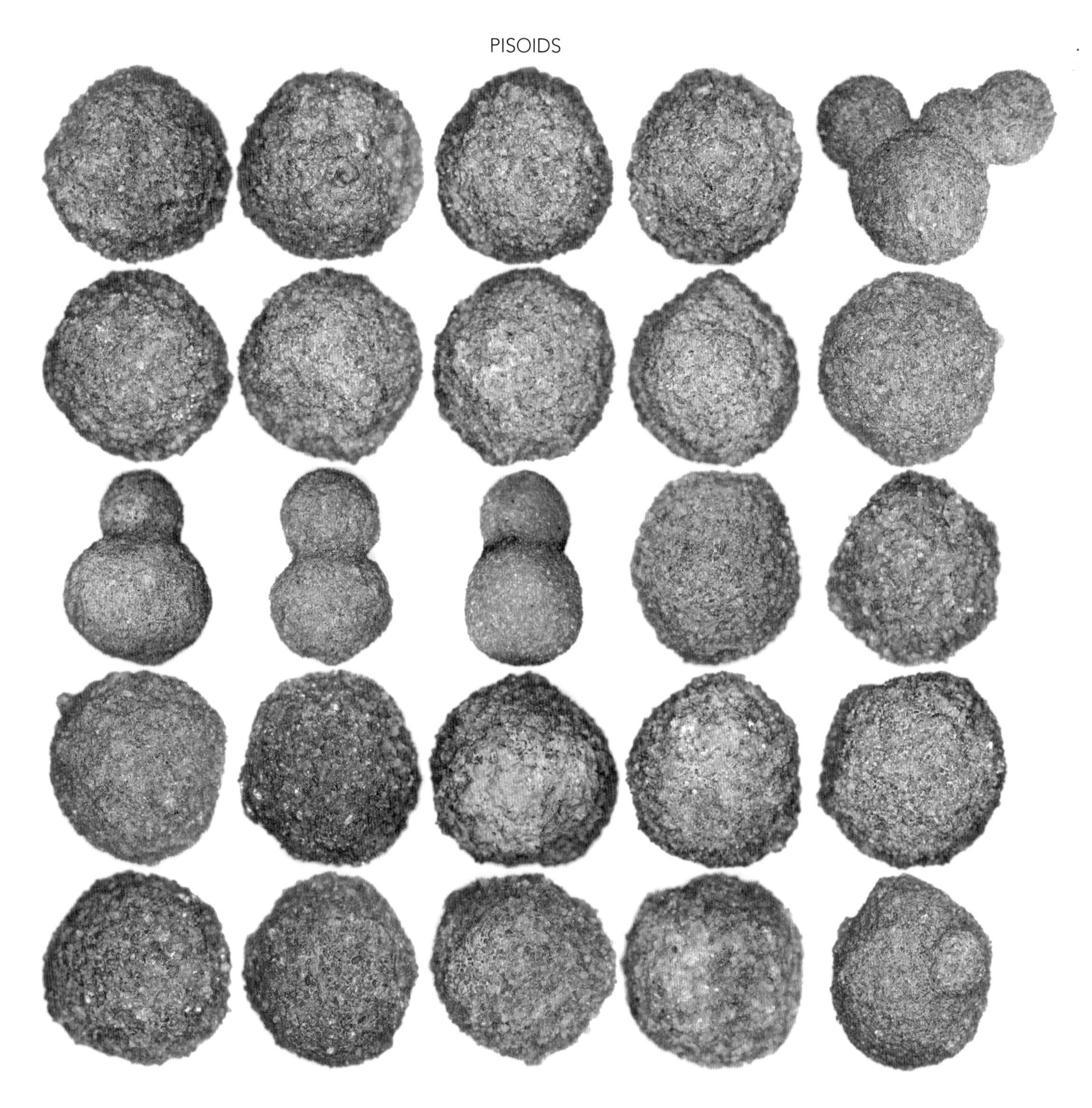

Pele's Tears—Achneliths

Among the many types of mineral dust being produced and moved around the globe by the wind (see page 123), volcanic tephra are the most dramatic. Throughout history entire civilizations have been buried, and today volcanic ash is causing problems to air traffic, etc. Some of the volcanic dust particles may at first glance look like S-type glassy micrometeorites, but can if necessary be distinguished from micrometeorites by a simple chemical analysis.

Pele's tears, or achneliths, are spherical pyroclasts–drops of volcanic glass thrown out during an eruption. The glassy particles are formed by quenching of magma spray and are typically lapilli-sized, 2–64 mm, which is 10–100 times larger than average micrometeorites.

Achneliths can reveal a great deal of information about the eruption. Examination of bubbles of gas and particles trapped within the tears can provide information about the composition of the magma chamber. The shape of the tears can provide an indication of the velocity of the eruption. The achneliths on these pages are only 0.3–1.5 mm, and were found on the crater rim of the vulcano Le Piton de la Fournaise, La Reunion in the Indian Ocean.

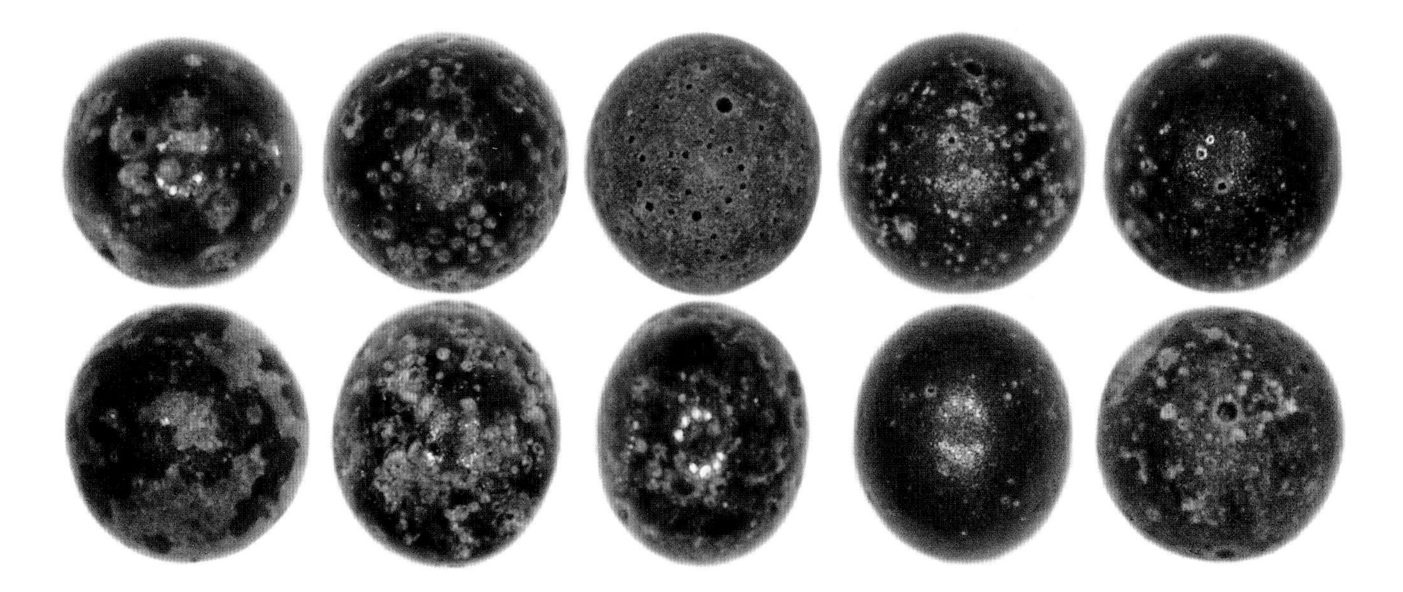

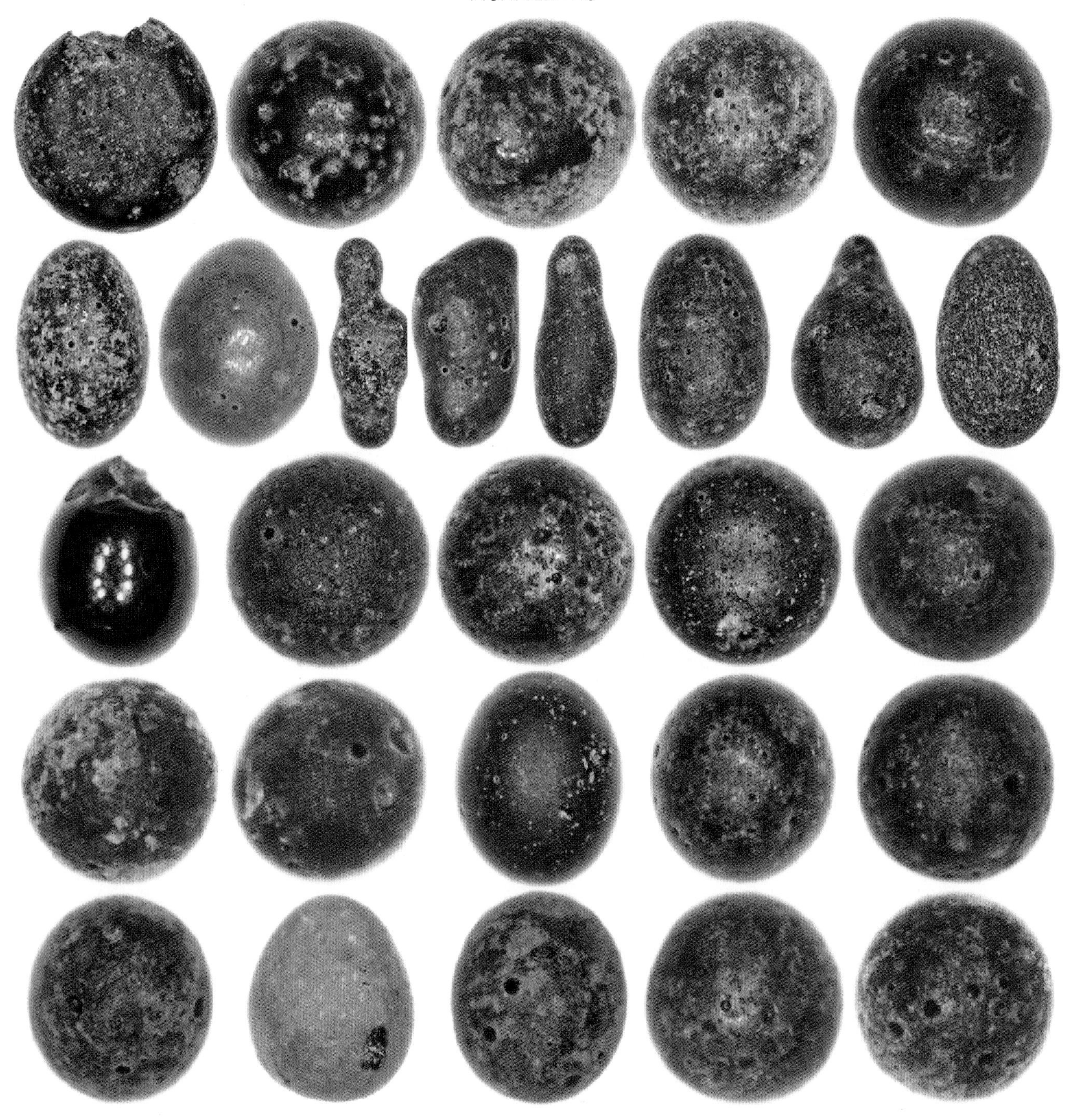

Road Dust Crystals

In the hunt for micrometeorites, while examining samples of road dust under the microscope we may encounter all sorts of objects—and occasionally nature's own jewelry—crystals. Without experience from the mineral kingdom one could suspect these amazing geometrical forms, colorful gems and shiny metal to be not of this world, but well developed crystals of this size indicates that they are of terrestrial origin.

Here is a collection of such submillimeter surprises found in common road dust while searching for micrometeorites—art by accident. They represent a previously undocumented mineralogical minimalism: erosion products and sediments of zero age in the urban landscape. See also the magnetite crystals on page 125. In pursuit of the extraterrestrial this is yet another category of particles to be aware of and disregard, but nonetheless to enjoy.

Thanks

Special thanks for SEM/eds analysis by the LABORATORY FOR ELECTRON MICROSCOPY at the University in Bergen, the NATURAL HISTORY MUSEUM in Oslo, the UNIVERSITY in OSLO (UiO), the INSTITUTE FOR ENERGY TECHNOLOGY (IFE), the NATURAL HISTORY MUSEUM (NHM) in London, OLYMPUS NORGE, OLYMPUS EUROPE, and all the people who have let me climb on their roofs to take dust samples.

These marvelous people, mentioned in alphabetical order, have contributed with content, analysis and/or inspiration for this book:

ALVE, Elisabeth
ANDENÆS, Ragnar
ANDERSON, Don
BERG, Berit Løken
BERGO, Torbjørn
BERGSTRØM, Arnulv
BESEDIN, Thaddeus
BETTINI, Giuliano
BILET, Morten
BRADBURY, David
BRATFOSS, Rune
COLOMBETTI, Alessandro
DUFTER, Alfred
DYPVIK, Henning
ERICHSEN, Egil Severin
FINSTAD, Hanne
GARVIE, Laurence
GENGE, Matthew
GILMER, Michael
GJELSVIK, Norvald
GLASS, Billy
GRANDE-JOHNSEN, Kent-Rune
GÜTHMANN, Michael
HABIBI, Aziz
HARPER, Johnnie
HEGGSTAD, Irene
JAHN, Barbara
JAKUBOWSKI, Tomasz
JOHANSEN, Annika
KARTASHOV, Pavel M.
KIHLE, Jan Braly
KOLDRUP, Harald
KUBALCZAK, Tomek
LINDHOLM, Markus
LINDSETH, Trond
MATHEWS, John D.
MAURETTE, Michel
NAKREM, Hans Arne
NILSEN, Odd
NOVAK, Martin J.
OLUFSEN, Kjell
PRACAS, Peter
PRASAD, Shyam M.
RYKKJE, Johannes
SCHALLER, Emily
SELBEKK, Rune S.
SINGH, Dupinder
SKUBLOV, Gennady
STANKIEWICZ, Przemyslaw
SÆLEN, Gunnar
TAYLOR, Susan
THORESEN, Øivind
VALEV, Ventsislav K.
WHYMARK, Aubrey

Tabula Gratulatoria

The book *In Search of Stardust* was first published with no other external funding other than the preorders from the following splendid persons, with special thanks to these:

Chris Watson, Niels Højgaard Andersen, Rok Gasparic, Patrick Brown, Jon Wallace, Inger Kjersti Iden, Jon Erik Eriksen, Andràs Fegyvàri, David Gonzales, Jonathan Kay, Jon Phillips, Nathan Swanson, Dolora Westrich, Craig Whitford, Arne de Gros Dich, Stefan Carlgren, Kristian Eek Haugen, Ljubomir Nestorovic, Tor Arne Holm, Andries Goedhart, Douglas Smith, Stein Rørvik, Marion Delannoy, Stephane Vermeulen, Eddie Johanna Dehls, Kalle Guldbrandsen, Hristijan Mitrevski, John Shea, Ragnhild Krogvig Karlsen, Arnulv Bergstrøm, Francesco Nicolodi, Runar Sandnes, Andre Moutinho, Martin Goff, Peter Bronton, Wenche Svindal, Erik Daems, Colin Cameron, Terje Fjeldheim, Norvald Gjelsvik, Roald Ellingsen, Elin Birgitte Sagvold, Helge Hustveit, Michael J. Simms, Nelson Holcomb, Grace Rivas Seland, Olav Bonifacio Rivas Seland, Eduardo Jawerbaum, Klaus Giesselmann, Torfinn Kjærnet, Espen Kolberg, Bethel Tzhaye, Johannes Fantayebil Kolberg, Lydia Fantayebil Kolberg, Tore Furuheim, Michael Hviid, Manfred Heising, Pierre Bels, Sam Crossley, Richard Zimmerman, Knut Edvard Larsen, Ole Tjugen, Jen Makowsky, Jan Strebel, Kieran Davis, Graham Ensor, Thomas Hughes, Zbigniew Godwinski, Nina Thomassen, Olav Rokne Erichsen, Gro Eileraas, Markus Lindholm, Øystein Johannessen, Taylor Trott, Daniel Wray, Ola Antonsen, Jacob Wilk, Rob Wesel, Bente Veronica Johansen, Jean-Marie Biets, Arild Sakshaug, Michael Santos, Sigbjørn Mork, Daniel Belliveau, Troy Bell, Thor Sørlie, Herløv Haug, Roy Magnus Andersen, Heidi Cathrin Størholt, Dale Nason, Pål Tore Mørkved, John Scott Parker, and Pawel Sikorski.

Index